MATHS IN ACTION

ADVANCED HIGHER

Mathematics 2

Edward Mullan
Peter Westwood
Clive Chambers

This edition first published in 2000 by Nelson Thornes Ltd

Reprinted in 2001 by:
Nelson Thornes Ltd
Delta Place
27 Bath Road
CHELTENHAM
GL53 7TH
United Kingdom

ISBN 978 0 17 431542 1
07 08 / 10 9 8 7 6

Typeset by Upstream, London
Printed and bound in Croatia by Zrinski d.d., Cakovec.

Acknowledgements
The authors and publisher are grateful for permission to include the following copyright material:
Corbis UK Ltd; p.13 (Bettman), p.15 (Bettman), p.27 (Michael Nicholson), p.48 (Leonard de Selva), p.59 left (Bettman), p.79 left (Bettman), p.79 right (Bettman), p.106: Hulton Getty Picture Collection Ltd; p.88, p.99: Michael Holford; p.113: Science and Society Picture Library; p.1: Science Photo Library; p.59 right: Zooid pictures; p.21.

Every effort has been made to trace all the copyright holders, but where this has not been possible the publisher will be pleased to make any necessary arrangements at the first opportunity.

Contents

Preface

This book is part of the *Maths in Action* series, and has been written to address the needs of students following Advanced Higher Mathematics, unit 2, Mathematics 2. It is assumed that students have already done Higher Mathematics and Advanced Higher, Mathematics 1, and each chapter contains some reminders of necessary knowledge.

The order of the chapters and the topics within chapters mirrors that in the detailed contents list of the national course specifications, with the exception of Chapter 1 on Proof and Elementary Number Theory, since the methods of proof are required for most of the rest of the book.

Chapter 2, on the large topic of Differentiation, has been split into two parts – theory and applications – to make it more manageable.

Notation and conventions follow those of the book *Higher Mathematics, Maths in Action*. The style also reflects earlier books in the Maths in Action series so that students can follow this course with as smooth a learning curve as possible. Some liberties have been taken in mixing language and notation to ease the student into more rigorous thought as gently as possible. Explanations have been kept short and to the point to ensure that the topics remain accessible.

Minor historical notes have been added at places throughout the text in the hope that students are made curious enough to pursue the topics on their own. The historical background to a piece of maths (dates, personalities, inspiration and development) often helps to put it in a context.

Each chapter ends with two features:

- a Review exercise which has the purpose of assuring the student that he or she has indeed picked up all the learning outcomes associated with that particular chapter;
- a Summary which provides the basis of revision notes.

The use of graphic calculators and/or spreadsheets is encouraged, though not essential, at many places. Being aware that variations occur between commercial packages, we have tried to describe approaches in generic terms where possible.

1 Proof and Elementary Number Theory

Historical note

Sophie Germain

Many mathematicians contributed to proof and number theory. One of the most noted is Sophie Germain (1776–1831).

She wrote several papers and letters to other mathematicians of her time under the pseudonym M. LeBlanc because she feared that, being a woman, she would be ignored.

Her most famous correspondence was with Gauss who, without knowing her, gave her proofs on number theory a lot of praise. He only came to know her true identity when she intervened on his behalf when the French occupied his town. She did a lot of important work on Fermat's last theorem.

Some definitions

A *statement* is any sentence which is either true or false but not both. For example

a:	$3 + 5 = 4$	false
b:	$4 + 5 > 10$	false
c:	$2x + 3x = 5x$	true

The *negation* of a statement can usually be constructed by putting 'not' after the verb of the sentence. When a statement a is true, its negation *not-a* will be false, and vice versa. For example

negation of a:	$3 + 5 \neq 4$	true
negation of b:	$4 + 5 \leq 10$	true
negation of c:	$2x + 3x \neq 5x$	false

A *compound statement* can be formed from given statements by combining them in various ways, for example by using

and:	$(3 + 5 = 4)$ and $(4 + 5 > 10)$	false
or:	$(3 + 5 = 4)$ or $(4 + 5 > 10)$	false
If...then:	If $(3 + 5 = 4)$ then $(4 + 5 > 10)$	true

A *universal statement* refers to all items in a set. For example

$2x + 3x = 5x$ for all $x \in R$

$x^2 + 1$ *is odd* for all $x \in Z$

If a universal statement is false, this can be proved by citing one *counterexample*. For example, the statement '$x^2 + 1$ *is odd* for all x' is false since, for $x = 1$, $x^2 + 1 = 2$, which is even.

An *existential statement* states that there exists at least one item in a set that has a special property. For example

There exists an x such that $2x + 3x \neq 5x$.
There exists an x such that $x^2 + 1$ is even.

If an existential statement is true, this can be proved by citing one *example*.

'For all x' is often denoted by '$\forall x$' and 'There exists an x' is often denoted by '$\exists x$'.

For all x, $2x + 3x = 5x$ is the same as $\forall x, 2x + 3x = 5x$
There exists an x such that $2x + 3x \neq 5x$
is the same as $\exists x$ such that (s.t.) $2x + 3x \neq 5x$ *or* $\exists x, 2x + 3x \neq 5x$

Related implications

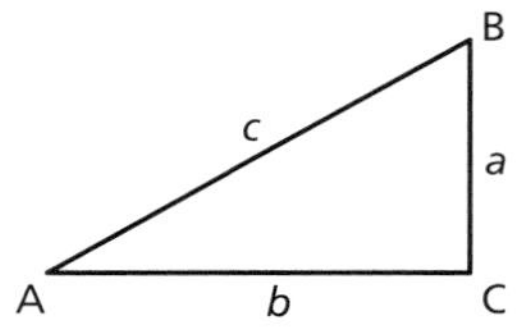

An *If...then* statement is often referred to as an *implication*.
For example, the theorem of Pythagoras is an implication.

If triangle ABC is right angled at C *then* $c^2 = a^2 + b^2$.

There are three related implications.

Its *inverse* is: *If* triangle ABC is not right angled at C *then* $c^2 \neq a^2 + b^2$.
Its *converse* is: *If* $c^2 = a^2 + b^2$ *then* the triangle ABC is right angled at C.
Its *contrapositive* is: *If* $c^2 \neq a^2 + b^2$ *then* the triangle ABC is not right angled at C.

Example Write down the inverse, converse and contrapositive of the statement '*If* the shape is a square *then* it has four sides.'

Its *inverse* is: *If* the shape is not a square *then* it does not have four sides.
Its *converse* is: *If* it has four sides *then* the shape is a square.
Its *contrapositive* is: *If* it does not have four sides *then* the shape is not a square.

Note
If the original implication is true then the contrapositive is true, but the inverse and converse need not be true.

The symbol '$\Rightarrow$' is used to denote '*implies*'. For example

The triangle ABC is right angled at C $\Rightarrow c^2 = a^2 + b^2$

is the same as

If the triangle ABC is right angled at C *then* $c^2 = a^2 + b^2$

When the converse is also true, i.e.

$c^2 = a^2 + b^2 \Rightarrow$ the triangle ABC is right angled at C

then we use the symbol '$\Leftrightarrow$' and write

the triangle ABC is right angled at C $\Leftrightarrow c^2 = a^2 + b^2$

The symbol $\Leftrightarrow$ is often read as '*if and only if*'. This is referred to as a *two-way implication* or *equivalence*.

EXERCISE 1A

1 **(i)** Write down the negation of each of the following statements.
(ii) State whether the negation is true or false.

a $3x + 5x = 7x$ **b** $3 + 6 > 7$ **c** $4^3 \geq 3^4$

d $-1 \leq \sin x \leq 1$ **e** $\dfrac{d}{dx} e^x = e^x$ **f** $2x + 1$ is odd given $x \in W$

2 The negation of the statement 'All grass is green' is 'Not all grass is green' or 'Some grass is not green.' Write down the negation of each of the following statements.

a For all whole numbers, $2x + 1 > 5$. **b** For all x, $\dfrac{1}{x}$ exists.

c $\forall x, x^2 > 0$ **d** $\forall n, 2^n + 1$ is prime

3 The negation of the statement 'Some grass is green' is 'All grass is not green.' State the negation of each of the following statements and say whether it is true or false.

a Some whole numbers of the form $2x + 1$ are even. **b** For some x, $2^x > 3^x$.

c There exists a natural number less than 1. **d** There exists an even prime.

e $\exists x$ such that $\ln x > \ln(x + 1)$ **f** $\exists x$ such that $\sqrt{x} \notin R$

4 For each implication, state **(i)** its inverse, **(ii)** its converse, **(iii)** its contrapositive.

a *If* a shape is an isosceles triangle *then* it will have two equal sides.

b *If* a whole number can be divided by 2 without remainder *then* it is even.

c *If* $2x + 4 > 12$ *then* $x > 4$. **d** $\sqrt{x} > 1 \Rightarrow x > 1$

5 The converse of an implication can be written using the symbol '$\Leftarrow$'. For example

$$2x + 1 = 5 \Rightarrow x = 2 \quad \text{has a converse} \quad 2x + 1 = 5 \Leftarrow x = 2$$

For the following implications **(i)** state the converse using '$\Leftarrow$',
(ii) say whether the implication and/or its converse is true,
(iii) hence say whether the implication is two-way.

a $2x + 1 = 5 \Rightarrow x = 2$ **b** $x = y \Rightarrow x + z = y + z$ **c** $x = y \Rightarrow xz = yz$

d $x = y \Rightarrow \dfrac{x}{z} = \dfrac{y}{z}, z \in W$ **e** $x = \dfrac{\pi}{6} \Rightarrow \sin x = \dfrac{1}{2}$ **f** $x = 1 \Rightarrow \ln x = 0$

g $x = 1 \Rightarrow x^2 = 1$ **h** x is even $\Rightarrow x^2$ is even **i** $y = x^3 \Rightarrow \dfrac{dy}{dx} = 3x^2$

6 Identify which of the following pairs of statements form true equivalences, and write them down using the symbol '$\Leftrightarrow$'.

a The shape is a triangle; the shape's internal angles sum to 180°.

b The shape is an isosceles triangle; the shape has one axis of symmetry.

c $2x^2 + 1 = 19$; $x = 3$ **d** $a \geq b$; $\ln a \geq \ln b$

e $a \geq b$; $\sin a \geq \sin b$ **f** $a \geq b$; $ac \geq bc$, $c \in W$

g $a \geq b$; $ac \geq bc$, $c \in R$ **h** $a \neq 0$ *and* $b \neq 0$; $ab \neq 0$

i $a = 0$ *or* $b = 0$; $ab = 0$ **j** $a = -b$ and $c = 0$; $a + b + c = 0$

7 Find a counterexample to disprove each of the following universal statements.

a $\forall x \in W, 5x > 4x$ **b** $\forall x \in R, 2x \neq 2^x$

c $\forall x \in R, ax = bx \Rightarrow a = b$ **d** $\forall x \in Z, \sqrt{x^2} = x$

e $\forall x \in R, \sqrt{1 - \sin^2 x} = \cos x$ **f** $\forall a, b \in R, |a| + |b| \leq |a + b|$

g $\forall x \in W, e^{\ln x} = x$

8 Find an example to prove each of the following existential statements.

a $\exists x \in W,\ 5x \leq 4x$

b $\exists x \in R,\ 2x = 2^x$

c $\exists a, b \in R,\ (a + b)^2 = a^2 + b^2$

d $\exists A, B \in R,\ \sin(A + B) = \sin A + \sin B$

e $\exists x \in R,\ \dfrac{x}{x} \neq 1$

f $\exists a, b \in R,\ |a| + |b| \leq |a + b|$

g $\exists x \in W,\ \tan^{-1}(\tan x) \neq x$

Direct proof

What is *proof*?

A proof is an argument in which the truth of a statement is established, in a logical set of steps, from the truth of a given statement or set of statements.

The given statement is either a statement whose truth is to be taken for granted (an *axiom*), or a statement whose truth has been established by a previous proof (a *theorem*).

In a *direct proof* we move directly from statements whose truth is given, by a series of implications, to the truth of another statement.

Direct proof framework

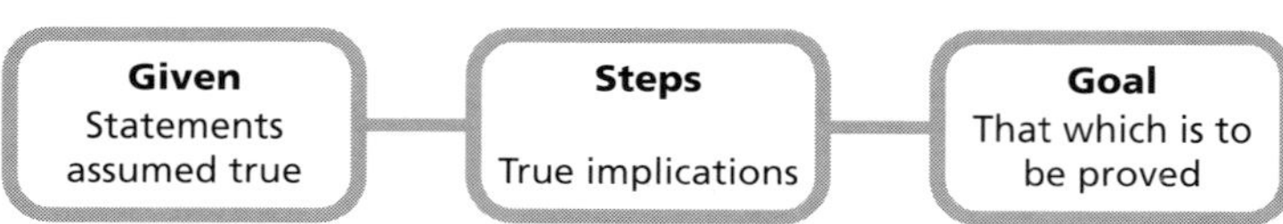

Example 1 Given that $5x - 6 = 2x$ prove that $x = 2$.

$5x - 6 = 2x \Rightarrow 3x - 6 = 0$

$3x - 6 = 0 \Rightarrow 3x = 6$

$3x = 6 \Rightarrow x = 2$

Thus $5x - 6 = 2x \Rightarrow x = 2$

$5x - 6 = 2x$ is true (given)

Thus $x = 2$ is true which is what we wanted to prove.

If we represent the statement $5x - 6 = 2x$ by a, the statement $3x - 6 = 0$ by b, the statement $3x = 6$ by c and the statement $x = 2$ by d, we see that the proof takes the form:

Given a prove d. *Given statement, and declared goal*

$a \Rightarrow b$

$b \Rightarrow c$

$c \Rightarrow d$

} *Steps: a sequence of true implications*

Thus $a \Rightarrow d$

a is true (given)

Thus d is true. *Goal achieved*

This then is the *shape* of any direct proof:

any number of *linked* implications leading to
given implies *goal*
and if *given* is true
this proves *goal* is true

Care must be taken to ensure the linkage of the implications is as shown, namely $a \Rightarrow b$, $b \Rightarrow c$, $c \Rightarrow d$, etc. The false linkage $a \Rightarrow b$, $c \Rightarrow b$ does not generally mean that $a \Rightarrow c$. For example

'$x = 10 \Rightarrow x$ is even', '$x = 20 \Rightarrow x$ is even' does not mean '$x = 10 \Rightarrow x = 20$'

Each implication in the set of steps must be true or the proof fails. For example, if the implication $x^2 = 49 \Rightarrow x = 7$ occurs, then the conclusions reached will only be valid for, say, $x \in W$.

Example 2 The following argument contains two errors. Find the steps that contain errors.

Given $t \in R$, $t = \sqrt{(3 - 2x + 4 + x)}$, prove $x > 7$.

$$t = \sqrt{(3 - 2x + 4 + x)} \Rightarrow t = \sqrt{(7 - x)} \qquad 1$$
$$t = \sqrt{(7 - x)} \Rightarrow 7 - x > 0 \qquad 2$$
$$7 - x > 0 \Rightarrow x > 7 \qquad 3$$

Given $t = \sqrt{(3 - 2x + 4 + x)}$ is true then $x > 7$ is true.

The first error occurs in line 2 where the true implication is $t = \sqrt{(7 - x)} \Rightarrow 7 - x \geq 0$

The second error occurs in line 3 where the implication should be $7 - x \geq 0 \Rightarrow x \leq 7$

The corrected proof, for convenience, can be written as

Given $t \in R$, $t = \sqrt{(3 - 2x + 4 + x)}$, prove $x \leq 7$.

$$t = \sqrt{(3 - 2x + 4 + x)} \Rightarrow t = \sqrt{(7 - x)}$$
$$\Rightarrow 7 - x \geq 0$$
$$\Rightarrow x \leq 7$$

Given $t = \sqrt{(3 - 2x + 4 + x)}$ is true then $x \leq 7$ is true.

EXERCISE 1B

1 Here are two well-known false arguments. Find the steps which contain errors.

a Given $x = y$, prove that $1 = 2$.

$$x = y \Rightarrow x^2 = y^2 = xy$$
$$\Rightarrow x^2 - xy = x^2 - y^2$$
$$\Rightarrow x(x - y) = (x - y)(x + y)$$
$$\Rightarrow x = x + y$$
$$\Rightarrow x = 2x \quad \text{(since } x = y\text{)}$$
$$\Rightarrow 1 = 2$$

Given $x = y$ is true then $1 = 2$ is true.

b Given $\frac{1}{2} > \frac{1}{4}$, prove that $1 > 2$.

$$\frac{1}{2} > \frac{1}{4} \Rightarrow \ln\left(\frac{1}{2}\right) > \ln\left(\frac{1}{4}\right)$$
$$\Rightarrow \ln\left(\frac{1}{2}\right) > \ln\left(\left(\frac{1}{2}\right)^2\right)$$
$$\Rightarrow \ln\left(\frac{1}{2}\right) > 2\ln\left(\frac{1}{2}\right)$$
$$\Rightarrow 1 > 2$$

Given $\frac{1}{2} > \frac{1}{4}$ is true then $1 > 2$ is true.

2 Read through each of these short *proofs*. If you think there is a fault, state the line number and why the line is wrong.

a Given $x \in Z$ *and* $x^2 + 3x = 3x + 16$, prove $x = 4$.

$x^2 + 3x = 3x + 16$	1
$\Rightarrow x^2 = 16$	2
$\Rightarrow x = 4$	3
If the given statements are true then $x = 4$ is true.	4

b Given ABCD is a parallelogram *and* BC = 7, CD = 6, BD = 11, prove AC = 11.

ABCD is a parallelogram $\Rightarrow$ AC = BD	1
$\Rightarrow$ AC = 11 (since BD = 11)	2
If the given statements are true then AC = 11 is true.	3

c Given PQ = 12, PR = 5, QR = 13, prove triangle PQR is right angled.

$12^2 = 144$	1
and $5^2 = 25$	2
$\Rightarrow 5^2 + 12^2 = 169$	3
$\Rightarrow PR^2 + PQ^2 = QR^2$ (since $13^2 = 169$)	4
$\Rightarrow$ triangle PQR is right angled at Q (by the converse of the theorem of Pythagoras)	5
If the given statements are true then triangle PQR is right angled is true.	6

d Given $4 - 7(x - 3) < 2x + 5$, prove $x > -4$.

$4 - 7(x - 3) < 2x + 5$	
$\Rightarrow 4 - 7x + 21 < 2x + 5$	1
$\Rightarrow -5x < 20$	2
$\Rightarrow x > -4$	3
If the given statements are true then $x > -4$ is true.	4

e Given that ln 8 and ln 3 are irrational and that $3^x = 8$, prove x is irrational.

$3^x = 8$	
$\Rightarrow x \ln 3 = \ln 8$	1
$\Rightarrow x = \dfrac{\ln 3}{\ln 8}$	2
$\Rightarrow x = \dfrac{\text{an irrational number}}{\text{an irrational number}}$	3
$\Rightarrow x$ is an irrational number	4
If the given statements are true then x is irrational is true.	5

3 Look at the short proofs below, and for each one write down the line number(s) where you would insert an implication sign ($\Rightarrow$).

a Given that $x^2 + 4x - 2 = 0$, prove $x = -2 \pm \sqrt{6}$.

$x^2 + 4x - 2 = 0$	1
$(x + 2)^2 - 4 - 2 = 0$	2
$(x + 2)^2 = 6$	3
$x + 2 = \pm\sqrt{6}$	4
$x = -2 \pm \sqrt{6}$	5

If the given statement is true then it is true that $x = -2 \pm \sqrt{6}$.

b Given x is an even number, prove x^2 is an even number.

x is an even number

$x = 2k$ where $k \in W$	1
$x^2 = (2k)^2$	2
$x^2 = 4k^2$	3
$x^2 = 2(2k^2)$	4
x^2 is an even number	5

If the given statement is true then it is true that x^2 is an even number.

c Given that AC = 15, AD = 9 *and* ABCD is a rectangle, prove DC = 12.

$AC^2 = 225$	1
$AD^2 = 81$	2
$DC^2 = AC^2 - AD^2$	3
$DC^2 = 225 - 81$	4
$DC^2 = 144$	5
$DC = 12$	6

If the given statements are true then it is true that DC = 12.

d Given $t = \dfrac{1}{2 - \sqrt{3}} - \sqrt{3}$, prove t is rational ($t \in Z$).

$t = \dfrac{1}{2 - \sqrt{3}} - \sqrt{3}$	1
$t = \dfrac{1}{2 - \sqrt{3}} \times \dfrac{2 + \sqrt{3}}{2 + \sqrt{3}} - \sqrt{3}$	2
$t = \dfrac{2 + \sqrt{3}}{4 - 3} - \sqrt{3}$	3
$t = 2 + \sqrt{3} - \sqrt{3}$	4
$t = 2$	5
t is rational since $2 \in Z$	6

If the given statement is true then it is true that t is rational.

4 Each part below consists of two implications (either of which may be true or false). If each pair of statements is true, then they may be replaced by a single equivalence. Either write down the equivalence or give a counterexample to prove one of the statements false.

a $x = 7 \Rightarrow x^2 = 49$
$x^2 = 49,\ x \in N \Rightarrow x = 7$

b $x = 7 \Rightarrow x^2 = 49$
$x^2 = 49,\ x \in R \Rightarrow x = 7$

c A, B, C are collinear $\Rightarrow \overrightarrow{AC} = \overrightarrow{AB} + \overrightarrow{BC}$
$\overrightarrow{AC} = \overrightarrow{AB} + \overrightarrow{BC} \Rightarrow$ A, B, C are collinear

d a and $b \notin Z \Rightarrow a + b \notin Z$
$a + b \notin Z \Rightarrow a$ and $b \notin Z$

e a and $b \notin Z \Rightarrow ab \notin Z$
$ab \notin Z \Rightarrow a$ and $b \notin Z$

f $3x + 3 = 15 \Rightarrow x = 4$
$x = 4 \Rightarrow 3x + 3 = 15$

5 By finding a counterexample, prove that each of the following conjectures is false.

a $u_n = n^3 - 6n^2 + 13n - 7,\ n \in N$, gives the sequence of odd numbers 1, 3, 5, …

b Diagonals of a quadrilateral are equal $\Rightarrow$ the quadrilateral is a square.

c June has five Thursdays $\Rightarrow$ the first of June is a Thursday

d (Note: $p|a$ reads as 'p divides a' and means p is a factor of a.)
$p|(a + b)$ and $p|(b + c) \Rightarrow p|(a + c)$

e $\int x^n dx = \frac{1}{n+1} x^{n+1} + c \ \forall n \in Z$

f $S(n)$ is defined as the sum of the divisors of n other than n itself, for example $S(10) = 1 + 2 + 5 = 8$. Conjecture: $S(n) < n$ for all $n \in N$.

g $p|ab$ and $p \nmid a \Rightarrow p|b$ where the symbol '$\nmid$' means 'does not divide'.
[example of the conjecture when true: $5|(4 \times 15)$ and $5 \nmid 4 \Rightarrow 5|15$.

6 A famous conjecture states that $2^k - 1$ is prime whenever k is prime. Show that $k = 11$ is the smallest counterexample to this conjecture.

Looking for the right steps

Study the given statements and data.

- They may suggest some standard algorithm such as Pythagoras' theorem, trigonometric rules, etc.
- If the number x is known to be even, it often helps to express it as $x = 2k,\ k \in W$. Likewise, if the number x is odd, it can help to express it as $x = 2k + 1,\ k \in W$.
- If $b|a$ then we can write $a = kb,\ k \in W$.

Study the required goal.

- It may be similar to other problems.
- Its form may be reminiscent of known formulae.
- If $b|a$ then we can write $a = kb,\ k \in W$.
- Consider intermediate goals.
- Can the goal be expressed in another way?

Sometimes the steps come as a moment of inspiration. It is said that Gauss, at the age of 10, produced the following proof without instruction.

Example 1 Prove that if $S(n)$ is the sum of the first n natural numbers then $S(n) = \frac{1}{2}n(n+1)$.

$$S(n) = 1 + 2 + 3 + \cdots + (n-1) + n \qquad \textit{(given)}$$
$$\Rightarrow \quad S(n) = n + (n-1) + \cdots + 3 + 2 + 1 \qquad \text{(reversing the order of terms)}$$

Adding, we get

$$2S(n) = (n+1) + (n+1) + \cdots + (n+1) + (n+1) \quad \text{for } n \text{ terms}$$
$$\Rightarrow \quad 2S(n) = n(n+1)$$
$$\Rightarrow \quad S(n) = \frac{1}{2}n(n+1)$$

The inspiration is to see that the reversal step will eradicate the non-specific nature of the '…'.

Example 2 Prove that $n^2 + 3n$ is divisible by 2 for all $n \in N$.
[Prove $2 \mid (n^2 + 3n)\ \forall n \in N$.]

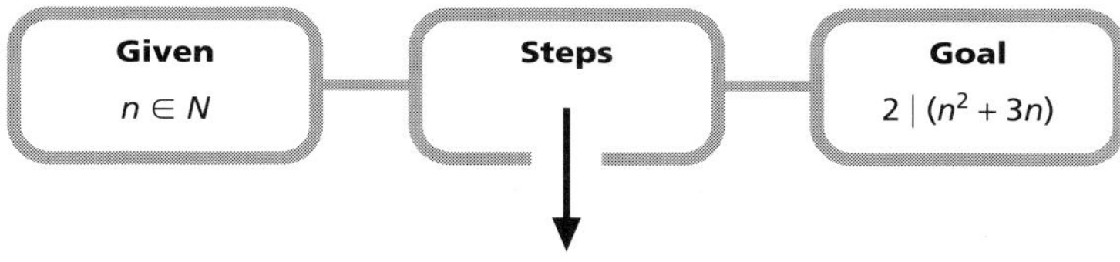

Thinking

What is given does not initially appear too helpful.
The goal suggests we want to prove $(n^2 + 3n) = 2a,\ a \in N$.
Factorising, we see that this means we wish to prove $n(n+3) = 2a$.
We can see that if n is even then it is true.
If n is odd then $n + 3$ is even and it is again true.

$n \in N \Rightarrow n$ is even or n is odd.

Case 1: n is even $\Rightarrow \quad n = 2k,\ k \in N$
$\Rightarrow \quad n^2 + 3n = (2k)^2 + 3(2k)$
$\Rightarrow \quad n^2 + 3n = 2(2k^2 + 3k)$
$\Rightarrow \quad n^2 + 3n = 2a,\ a \in N \quad$ (since $(2k^2 + 3k) \in N$)
$\Rightarrow \quad 2 \mid (n^2 + 3n)\ \forall$ even n

Case 2: n is odd $\Rightarrow \quad n = 2k - 1,\ k \in N \quad$ (If we express n as $2k + 1$, $k \in N$, we miss out the case when $n = 1$.)
$\Rightarrow \quad n^2 + 3n = (2k-1)^2 + 3(2k-1)$
$\Rightarrow \quad n^2 + 3n = 2(2k^2 + k - 1)$
$\Rightarrow \quad n^2 + 3n = 2a,\ a \in N \quad$ (since $(2k^2 + k - 1) \in N$)
$\Rightarrow \quad 2 \mid (n^2 + 3n)\ \forall$ odd n

$2 \mid (n^2 + 3n)\ \forall$ even n *and* $2 \mid (n^2 + 3n)\ \forall$ odd $n \Rightarrow 2 \mid (n^2 + 3n)\ \forall n \in N$

Example 3 Prove the sum S of $4 + 16 + 64 + \cdots + 4^n = \frac{4}{3}(4^n - 1)$.

$$S = 4 + 16 + 64 + \cdots + 4^n$$
$$\Rightarrow \quad 4S = 16 + 64 + \cdots + 4^n + 4^{n+1} \qquad \text{(multiplying throughout by 4)}$$
$$\Rightarrow \quad 3S = 4^{n+1} - 4 \qquad \text{(subtracting)}$$
$$\Rightarrow \quad S = \frac{4}{3}(4^n - 1)$$

EXERCISE 2A

1 Use the method of Example 1 above to prove the following.

a $2 + 4 + 6 + \cdots + 2n = n(n + 1)$ where $n \in W$

b $1 + 3 + 5 + \cdots + (2n - 1) = n^2$ where $n \in W$

2 Use the method of Example 3 to prove the following.

a $3 + 9 + 27 + 81 + \cdots + 3^n = \frac{3}{2}(3^n - 1)$ where $n \in W$

b $4 + 12 + 36 + \cdots + 4 \times 3^{n-1} = 2(3^n - 1)$ where $n \in W$

3 When a whole number is divided by 2 there are only two possibilities: either it divides exactly or there is a remainder of 1. So any whole number can be written in the form $2k$ or $2k + 1$ where $k \in W$.

a If $k \in N$ then these forms would look slightly different. How would they look?

b Prove that $n(n + 1)$ is always divisible by 2 where $n \in N$.

c When a natural number is divided by 3 there are only three possibilities: either it divides exactly or there is a remainder of 1 or there is a remainder of 2.

(i) Write down these three possibilities in terms of $k \in W$.

(ii) Prove that $n(n + 1)(n + 2)$ is divisible by 3 where $n \in W$.

d By considering parts **b** and **c** prove that the product of three consecutive numbers is divisible by 6.
[Hint: express the three numbers as n, $(n + 1)$ and $(n + 2)$.]

4 Prove that $n^3 - n$ is always divisible by 6.

5 Three edges of a tetrahedron meet at a vertex so that a right angle is formed between each pair of edges. Prove that the base triangle cannot be right angled.

6 This is an example of a classic problem: the stamp problem. In my stamp box I only have 2p stamps and 5p stamps. By considering even amounts and odd amounts, prove that I can post packages which require stamps of any total value of 4p or more.

7 Conjecture: 'If n is odd, then $n^2 + 1$ is even.'
Give a direct proof of this conjecture. A starting point could be 'n is odd $\Rightarrow n = \ldots$'.

8 Prove that $\sin x = k \sin(x + 2y) \Rightarrow \tan(x + y) = 1 + \frac{1 + k}{1 - k} \tan y$, where $k \neq 1$ and $x, y, k, \in R$.
[Hint: proofs depend on previous knowledge and spotting a starting point.
Previous knowledge: $\tan A = \frac{\sin A}{\cos A}$ and $\sin(A \pm B) = \sin A \cos B \pm \cos A \sin B$.
Starting point: express $(x + 2y)$ as $((x + y) + y)$ and x as $((x + y) - y)$.]

EXERCISE 2B

1 By considering the *units digit* prove that $6^n + 4$ is always divisible by 10 where $n \in N$.

2 By considering the factors of $3^{2n} - 1$ prove that $3^{2n} + 7$ is always divisible by 8 where $n \in N$.

3 Prove that $9 \times 1\,099\,999\,999\,989 = 9\,899\,999\,999\,901$.
[Hint: start by expressing 1 099 999 999 989 as the difference between two more convenient numbers and consider the expansion of the left-hand side.]

4 Prove that the sum of the squares of three consecutive integers cannot end in a 1, 3, 6 or 8.
[Hint: let the three consecutive integers be $x - 1$, x and $x + 1$, and try all 10 cases. For example, if x has a units digit of 2 then $5x^2 + 1$ will have a units digit of 1. This is often referred to as a proof by exhaustion.]

5 **a** What is the remainder when 5^3 is divided by 7?
b What is the remainder when 5^{99} is divided by 7?
[Hint: use part **a** and the binomial theorem.]

6 If n is an odd integer, prove that $n^2 - 1$ is divisible by 8.

7 **a** Prove that the product of three consecutive numbers plus the middle number is always a perfect cube.
b From your proof, write down a stronger conjecture than that given in part **a**.

8 The arithmetic mean of two numbers $a, b \geq 0$ is defined as $\frac{a + b}{2}$.
The geometric mean of two numbers $a, b \geq 0$ is defined as $\sqrt{(ab)}$.
Prove that, $\forall a, b \geq 0$, $\frac{a + b}{2} \geq \sqrt{(ab)}$.
[Hint: like many direct proofs it is not always obvious where to start.
Try starting with '$(a - b)^2 \geq 0$'.]
Note: this is a special case of the general theorem which states that for $a_1, a_2, \cdots, a_n \geq 0$

$$\frac{a_1 + a_2 + \cdots + a_n}{n} \geq \sqrt{a_1 a_2 \ldots a_n}$$

9 **a** Prove that $k(k^2 + 5)$ is divisible by 6.
b Hence prove that, if n is even, then $n^2(n^2 + 20)$ is divisible by 48.

10 Prove that $\cos^2 x + \cos^2\left(x + \frac{\pi}{3}\right) + \cos^2\left(x + \frac{2\pi}{3}\right) = \frac{3}{2}$.

Indirect proof

Proof by contradiction

Indirect proof is often referred to by the Latin expression *reductio ad absurdum* (reduce to an absurdity) because it has the following structure.

- Whatever statement we wish to prove, we assume its *negation* to be true.
- By a series of steps (valid implications) we arrive at some contradiction.
- Since all the steps are valid, it can only be the assumption which is false.
- If the negation is false, the original statement must be true.

Example 1 The ten digits 0, …, 9 can be arranged to form numbers which can then be summed. Each digit can be used once and only once in the formation of the numbers, for example

$$10 + 23 + 45 + 6 + 7 + 8 + 9 = 108$$

or $$20 + 38 + 19 + 4 + 5 + 6 + 7 = 99$$

Prove that it is impossible to arrange the ten digits into numbers whose sum is 100.

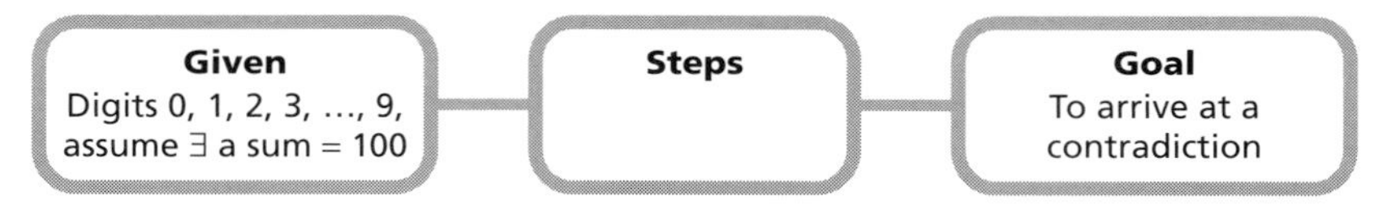

Assume there exists an arrangement of the digits which result in a sum of 100.

The sum of the digits is $0 + 1 + 2 + 3 + \cdots + 8 + 9 = 45$

Some of the digits will be used as *units* and the rest will be used as *tens*.

Let the whole number u represent the sum of the units digits. (Note: $u \in W$.)

Then $45 - u$ is the sum of the tens digits.

There exists an arrangement of the digits which result in a sum of 100.

$\Rightarrow \quad 100 = u + 10(45 - u) \;\Rightarrow\; 9u = 350 \;\Rightarrow\; u = \frac{350}{9}$

$\Rightarrow \quad u \notin W$ a contradiction

$\Rightarrow$ The assumption is false.

$\Rightarrow$ It is impossible to arrange the ten digits into numbers whose sum is 100.

Example 2 Prove $\frac{a+b}{2} \geq \sqrt{ab} \quad \forall a, b \in N$.

Assume $\frac{a+b}{2} < \sqrt{ab}$ for some $a, b \in N$

$\Rightarrow \quad a + b < 2\sqrt{ab}$

$\Rightarrow \quad (a+b)^2 < 4ab$ valid while $a, b \in N$

$\Rightarrow \quad a^2 + 2ab + b^2 < 4ab \;\Rightarrow\; a^2 - 2ab + b^2 < 0$

$\Rightarrow \quad (a-b)^2 < 0 \;\Rightarrow\; a, b \notin N$

$\Rightarrow$ The assumption is false.

$\Rightarrow \quad \frac{a+b}{2} \geq \sqrt{ab} \quad \forall a, b \in N$

Here is the proof devised by Euclid around 300 BC.

Euclid

Example 3 Prove that the number of prime numbers is infinite.

Assume that the number of primes is not infinite. Thus there exists a highest prime, p_n.

Let the list of *all* prime numbers be denoted by

$$p_1, p_2, p_3, \ldots, p_n$$

There exists a number $q \in N$ such that

$$q = p_1 \times p_2 \times p_3 \times \ldots \times p_n + 1$$

Clearly $q > 1$. Now

$$q \div p_1 = p_2 \times p_3 \times \ldots \times p_n \quad \text{remainder 1}$$

So q is not divisible by p_1.

Similarly q is not divisible by any of $p_2, p_3, \ldots, p_n$. But all natural numbers greater than 1 can be divided by 1 and at least one prime number. Thus a prime number exists which is not on the list.

We have a contradiction.

The number of prime numbers is infinite.

Example 4 Prove that $\sqrt{2}$ is not rational, that is $\sqrt{2} \notin Q$.

Assume that $\sqrt{2} \in Q$.

$\Rightarrow \quad \exists a, b \in Z,$ *where a and b have no common factors*, such that $\sqrt{2} = \dfrac{a}{b}$

$\Rightarrow \quad a = b\sqrt{2}$

$\Rightarrow \quad a^2 = 2b^2$

$\Rightarrow \quad a^2$ is even

$\Rightarrow \quad a$ is even, so a has a factor of 2

Let $a = 2k$ where $k \in W$.

Then $\quad 4k^2 = 2b^2 \quad$ (since $a^2 = 2b^2$)

$\Rightarrow \quad b^2 = 2k^2$

$\Rightarrow \quad b^2$ is even

$\Rightarrow \quad b$ is even, so b has a factor of 2

Both a and b are even.

$\Rightarrow \quad a$ and b have a common factor of 2.

We have a contradiction.

Thus $\sqrt{2}$ is not rational.

Proof by contrapositive

This type of proof depends on the fact that an implication and its contrapositive are logically equivalent. It is very useful when trying to prove the negation of a statement.

Example Prove that if $x = 77$ then x is not even.

'If x is even then x can be divided by 2 without remainder' — a definition

$\Leftrightarrow$ 'If x cannot be divided by 2 then x is not even' — the contrapositive

77 cannot be divided by 2.

$\Rightarrow$ 77 is not even

$\Rightarrow$ x is not even

EXERCISE 3A

1 Prove by contradiction that $\sqrt{3}$ is irrational (i.e. $\sqrt{3} \notin Q$).
[Hint: start by assuming the negation of this statement, $\sqrt{3} \in Q$, and remember that positive rational numbers can be written in the form $\frac{a}{b}$ where $a, b \in N$.]

2 Prove by contradiction that $\sqrt{7}$ is irrational.

3 Prove that $\frac{1}{2}(6\sqrt{3} - 1)$ is irrational.

4 Prove that no integers m and n can be found to satisfy the equation $14m + 20n = 101$.
[Hint: assume that they do exist and consider factors of both sides.]

5 Prove that no integers m and n can be found to satisfy the equation $14m + 21n = 100$.

6 *Conjecture*: If n is odd, then $n^2 + 1$ is even. Give a proof by contradiction of this conjecture.
[Hint: the negation of '$n^2 + 1$ is even' is '$n^2 + 1$ is not even' ($\Leftrightarrow$ '$n^2 + 1$ is odd'). A suitable starting point is 'Assume $n^2 + 1$ is not even $\Rightarrow n^2$ is ...'.]

7 Use proof by contrapositive to prove that 75 is an odd number.

8 **a** Prove by contradiction that if x^2 is even then x is even.
b Prove by contrapositive that if x is odd then x^2 is odd.

9 Use an indirect proof to show that, given $n \in Z$, $n^2 + n$ is even.

10 *Conjecture*: Given $a, b \in N$, if $a + b\sqrt{2} = 3 + 4\sqrt{2}$ then $a = 3$ and $b = 4$. Give an indirect proof of this conjecture.

11 Prove by contradiction that if the sum of two real numbers is irrational then at least one of the numbers must be irrational.

12 Use an indirect proof to show that, given $a, b \in Z$, if ab is even then at least one of a or b is even.

13 If two straight lines on a plane do not intersect then they are parallel. Prove by indirect means that if two straight lines are both perpendicular to a third line on the same plane then the two lines are parallel.

Aristotle

Aristotle stated that every demonstrable (i.e. provable) science must start from somewhere, namely the indemonstrable principles.

As we have seen, the proof of any statement depends on the truth of other, *given* statements. These, in turn, may have been proved by assuming the truth of more fundamental statements.

At some point we will reach the point where we need to say that 'this is self-evident' or 'this is obvious'. We come to statements which are accepted as true. We call such statements *axioms*.

The validity of every branch of mathematics depends on a set of axioms. Number theory is no different. We have already stated that saying that a number is of the form $2k$, $k \in W$, is equivalent to saying that the number is even; and saying that a number is of the form $2k + 1$, $k \in W$, is equivalent to saying that the number is odd. We have assumed that both these statements are self-evident.

But, in building up a sound foundation, we should not accept something as *axiomatic* if it can be proved. The second statement can be proved from the first.

Conjecture: All numbers of the form $2k + 1$ are odd numbers ($k \in W$).

Suppose $a = 2k + 1$ is even.

a is even $\Rightarrow a = 2p,\ p \in W$

$a = 2k + 1$ is even $\Rightarrow 2p = 2k + 1$ note that this implies $p > k$ and so $(p - k) \in W$

$\Rightarrow 2(p - k) = 1$

$\Rightarrow p - k = \frac{1}{2}$

But $(p - k) \in W$: contradiction.

So numbers of the form $2k + 1$ are odd. $a = 2k + 1 \Rightarrow a$ is odd

Conjecture: All odd numbers are of the form $2k + 1$ ($k \in W$).

Let b be an odd number.

b is odd $\Rightarrow b = a + 1$ where a is an even number

a is even $\Rightarrow a = 2k,\ k \in W$

$b = a + 1,\ a$ even $\Rightarrow b = 2k + 1,\ k \in W$

All odd numbers are of the form $2k + 1$. a is odd $\Rightarrow a = 2k + 1$

Bringing the two conclusions together proves the equivalence: a is odd $\Leftrightarrow a = 2k + 1$

The fundamental theorem of arithmetic

The statement known as the *fundamental theorem of arithmetic* was first proved in the nineteenth century by Gauss. Its proof depends on some earlier statements expounded in Euclid's *Elements* about 300 BC.

1 p divides a means p divides a without remainder (*definition*).

2 A prime number p is an integer greater than 1, which has only 1 and p as factors (*definition*).

3 If a prime number p divides ab, where a and b are integers, then either p divides a or p divides b.

3b If a prime number p divides ab, where a and b are primes, then $p = a$ or $p = b$.

4 Every integer greater than 1 can be written as a product of prime numbers.

The fundamental theorem of arithmetic develops this last statement:

Every integer greater than 1 can be written as a product of prime numbers *in only one way.*

Proof

Assume that a number k can be expressed as a product of primes in two ways:

$$k = p_1p_2p_3 \ldots \quad \text{and} \quad k = q_1q_2q_3 \ldots$$

So

$$p_1p_2p_3 \ldots = q_1q_2q_3 \ldots$$

After dividing throughout by any common prime factors we will be left with an expression of the form $p_ap_bp_c \ldots = q_aq_bq_c \ldots$

Thus $p_ap_bp_c \ldots$ and $q_aq_bq_c \ldots$ have no common prime factors.

But if p_a divides the left-hand side (LHS) it must also divide the right-hand side (RHS) since they are equal. If p_a divides $q_aq_bq_c \ldots$ then, by statement 3b above, it must equal one of the primes q_i.

Thus there is another common prime factor: contradiction.

So the prime factorisation of a must be unique.

This theorem is also referred to as the *unique factorisation theorem.* It is a very powerful theorem and the validity of much that you assume and use depends on it.

Example Prove that $\sqrt{2}$ is irrational using the fundamental theorem of arithmetic.

Assume $\sqrt{2} \in Q$.

$$\sqrt{2} \in Q \Rightarrow \sqrt{2} = \frac{a}{b} \text{ where } a, b \in Z$$

$$\Rightarrow a^2 = 2b^2$$

Let $2, p_2, p_3, \ldots, p_r$ be the combined list of distinct prime factors of a and of b.

Then a and b can be expressed as $\left(2^{k_1}p_2^{k_2}p_3^{k_3}\ldots p_r^{k_r}\right)$ and $\left(2^{l_1}p_2^{l_2}p_3^{l_3}\ldots p_r^{l_r}\right)$ respectively where $k_i, l_i \in W$.

$$a^2 = 2b^2 \Rightarrow \left(2^{k_1}p_2^{k_2}p_3^{k_3}\ldots p_r^{k_r}\right)^2 = 2\left(2^{l_1}p_2^{l_2}p_3^{l_3}\ldots p_r^{l_r}\right)^2$$
$$\Rightarrow 2^{2k_1}p_2^{2k_2}p_3^{2k_3}\ldots p_r^{2k_r} = 2^{2l_1+1}p_2^{2l_2}p_3^{2l_3}\ldots p_r^{2l_r}$$

By the fundamental theorem of arithmetic, these prime factorisations must be the same. In particular the factors of 2 must be identical.

$$2^{2k_1} = 2^{2l_1+1}$$
$$\Rightarrow 2k_1 = 2l_1 + 1$$
$\Rightarrow$ an even number is odd: contradiction

Thus $\sqrt{2} \notin Q$.

EXERCISE 3B

1 Use the fundamental theorem of arithmetic to prove by contradiction that $\sqrt{5}$ is irrational.

2 **a** Show by contradiction that if $\sqrt{ab}$ is irrational then $\sqrt{a} + \sqrt{b}$ is irrational, where $a, b \in N$.
[Hint: here we have to determine the truth of a compound statement which states '$\sqrt{ab}$ is irrational' $\Rightarrow$ '$\sqrt{a} + \sqrt{b}$ is irrational'. So, assuming the negation, start with '$\sqrt{a} + \sqrt{b}$ is rational' and use '$\sqrt{ab}$ is irrational' at some point to arrive at a contradiction.]

b Find a counterexample to the statement that

if $\sqrt{a} + \sqrt{b}$ is irrational then $\sqrt{ab}$ is irrational, $\forall a, b \in N$

3 Given that $|a| = \begin{cases} a & \text{when } a \geq 0 \\ -a & \text{when } a < 0 \end{cases}$ the statement $|a + b| \leq |a| + |b|$ is known as *the triangle inequality*. Prove the triangle inequality using proof by contradiction. [Hint: start by assuming the negation, and square both sides: $|a|^2 = a^2$.]

4 **a** Show that if $13x + 4$, $x \in W$, is a perfect square then $\exists a \in W$ such that $13x = (a - 2)(a + 2)$.

b Use the fundamental theorem of arithmetic to help you deduce that there is at least one prime number p for which $13p + 4$ is a perfect square.

5 **a** By considering the fundamental theorem of arithmetic find a prime number p such that $19p + 81$ is a perfect square.

b Find two non-prime whole numbers x which make $19x + 81$ a perfect square.

6 Find a prime number p so that $7p - 3$ is the product of two whole numbers which differ by 4.

7 **a** Identify a prime number p such that $7p + 1$ is a perfect cube.

b Find two non-prime integers x for which $7x + 1$ is a perfect cube.

Proof by induction

In many investigative circumstances you will have seen a repetitive process generate a sequence of numbers. A conjecture is made as to the formula for the nth term of the sequence, and the veracity of the conjecture is strengthened by considering further cases. No matter how many cases are considered, however, this does not constitute a proof.

The principle of *induction* provides a mechanism to prove such a conjecture.

1 A general statement is made about $n \in Z$, and demonstrated to be true for some value $n = a$.

2 It is then proved that, if the statement is true for $n = k$, then it is true for $n = k + 1$.

3 We can conclude that since it is true for $n = a$ (by step 1)
then it is true for $n = a + 1$ (by step 2)
and it is true for $n = a + 2$ (by step 2)
and so on

Thus, by induction, it is true for all $n \geq a, \quad n \in Z$.

Example 1 Consider the *triangular* numbers: the sequence is generated according to the recurrence relation $u_{n+1} = u_n + n + 1$

   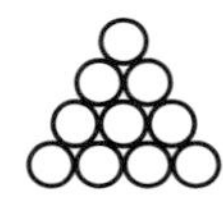 giving the sequence 1, 3, 6, 10, …

It is relatively easy to conjecture that the nth term is $u_n = \frac{1}{2}n(n + 1)$, $n \in N$. Prove this by induction.

1. We can see that the statement is true for $n = 1$: $u_1 = \frac{1}{2} \times 1 \times (1 + 1) = 1$

2. Assume that it is true for some value of $n = k$.
Thus

$$u_k = \frac{1}{2}k(k + 1)$$

> The goal is identified by replacing k by $k + 1$.
> $u_{k+1} = \frac{1}{2}(k + 1)((k + 1) + 1)$

Generate the next term using the recurrence relation.

$$u_{k+1} = \frac{1}{2}k(k + 1) + (k + 1) \qquad u_{n+1} = u_n + n + 1$$

$$\Rightarrow \quad u_{k+1} = (k + 1)(\tfrac{1}{2}k + 1)$$

$$\Rightarrow \quad u_{k+1} = \frac{1}{2}(k + 1)(k + 2)$$

$$\Rightarrow \quad u_{k+1} = \frac{1}{2}(k + 1)((k + 1) + 1)$$ arrange it in the form $u_n = \frac{1}{2}n(n + 1)$

$\Rightarrow$ the statement is true for $k + 1$

3. Since the statement is true for $n = 1$ and since (true for $n = k$) $\Rightarrow$ (true for $n = k + 1$) then, by induction, it is true $\forall n \geq 1$, $n \in N$.

Example 2 Prove that $n^3 + 2n$ is divisible by 3 $\forall n \geq 1, n \in N$.

1. The statement is true for the lowest required value of n, namely $n = 1$:
   $1^3 + 2 \times 1 = 3$, which is divisible by 3
2. Assume that it is true for some value $n = k$.
   Thus

   $k^3 + 2k = 3m$ for some $m \in N$

   > The goal is identified by replacing k by $k + 1$.
   > $(k + 1)^3 + 2(k + 1) = 3m_1$ for some $m_1 \in N$

   Consider $n = k + 1$.
   $$\begin{aligned}(k+1)^3 + 2(k+1) &= k^3 + 3k^2 + 3k + 1 + 2k + 2 \\ &= k^3 + 2k + 3k^2 + 3k + 3 \\ &= (k^3 + 2k) + 3(k^2 + k + 1) \\ &= 3m + 3(k^2 + k + 1) \quad \text{here we make use of the assumption that it is true for } n = k \\ &= 3(m + k^2 + k + 1) \\ &= 3m_1 \quad \text{for some } m_1 \in N\end{aligned}$$
   $\Rightarrow$ the statement is true for $k + 1$
3. Since the statement is true for $n = 1$ and since (true for $n = k$) $\Rightarrow$ (true for $n = k + 1$) then, by induction, it is true $\forall n \geq 1, n \in N$.

Example 3 Prove that $2^n > n, \forall n \in N$.

1. The statement is true when $n = 1$: $2^1 > 1$.
2. Assume it is true when $n = k$.
   Thus

   $2^k > k$

   > The goal?
   > To get to $2^{k+1} > k + 1$

   $\Rightarrow \quad 2 \times 2^k > 2k$ multiplying throughout by 2
   $\Rightarrow \quad 2^{k+1} > k + k$
   $\Rightarrow \quad 2^{k+1} > k + 1$ since $k \geq 1$
   $\Rightarrow$ the statement it true for $n = k + 1$
3. Since the statement is true for $n = 1$ and since (true for $n = k$) $\Rightarrow$ (true for $n = k + 1$) then, by induction, it is true $\forall n \geq 1, n \in N$.

EXERCISE 4

1 A recurrence relation, $u_{n+1} = u_n + 2n + 2$, $u_1 = 1$, generates the sequence 1, 5, 11, ... Prove by induction that the nth term of the sequence is $u_n = n^2 + n - 1$.

2 Prove that the recurrence relation defined by $u_{n+1} = u_n + 4n + 4$, $u_1 = 4$, has an nth term, $u_n = 2n(n + 1)$.

3 A particular linear recurrence relation is defined by $u_{n+1} = 2u_n + 1$, $u_1 = 2$. Prove that $u_n = 3 \times 2^{n-1} - 1$

4 Examine the number pattern being generated below.

$1 = 1^2$

$1 + 3 = 2^2$

$1 + 3 + 5 = 3^2$

$1 + 3 + 5 + 7 = 4^2$

a Make a conjecture about the sum of the first n odd numbers.

b Prove your conjecture by induction.

5

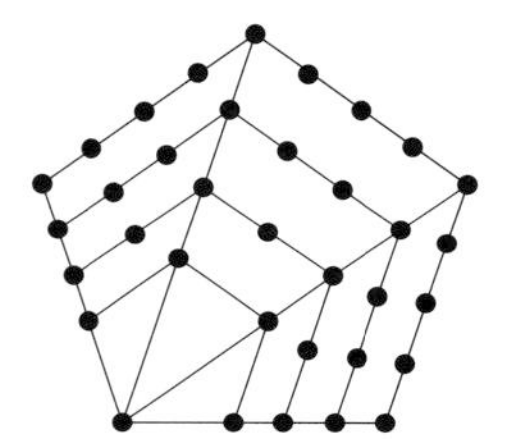

The *pentagonal* number pattern is generated by considering the number of points generated in a diagram similar to that shown, as the number of pentagons increases. Note the number of points per side increases by 1 each time. We get 1, 5, 12, 22, ... It is easy to see that the addition of the nth pentagon will add $3(n - 1) + 1$ points to the total.

a Write down the recurrence relation which generates the sequence.

b It has been conjectured that the formula for the nth pentagonal number is

$$u_n = \frac{n(3n - 1)}{2}.$$

Prove this by induction.

c Hexagonal numbers are generated in a similar fashion by considering hexagons. Prove that the nth hexagonal number is $n(2n - 1)$.

6 Prove by induction that

a $2^n > n, \quad \forall n \in N$

b $3^n > 2^n, \quad \forall n \in N$

c $2^n > n^2, \quad \forall n > 4, n \in N$

7 **a** Prove by induction that $5 + 7 + 9 + \cdots + (2n - 1) = n^2 - 4, \forall n \geq 3, n \in N$.

b Try forming a direct proof. This is shorter but not so easy to start.

8 Prove by induction that $2^{3n} - 1$ is divisible by 7, $\forall n \in N$.

9 **a** Prove by induction that $3^{2n} - 5$ is divisible by 4, $\forall n \in N$.
b Form a direct proof that 9^n is 1 more than a multiple of 8, i.e. $9^n = 8k + 1$, $n, k \in N$.
c **(i)** Hence prove directly that $3^{2n} - 5$, $n \in N$, is always divisible by 4.
(ii) Compare the length of working with that of part **a**.

10 The sequence 1, 1, 2, 3, 5, 8, 13, ... is called the *Fibonacci sequence* and is defined by the recurrence relation

$$F_{n+2} = F_{n+1} + F_n, \quad F_1 = F_2 = 1$$

The sequence first appeared in a problem in the book *Liber Abaci*, written by Leonardo De Pisa, also known as Fibonacci, in 1202. It was the Frenchman Edouard Lucas who attached Fibonacci's name to the sequence over 600 years later. The sequence can be found in many contexts: Pascal's triangle and the Golden Ratio to name but two.

Fibonacci

Prove the following results by induction.
a Sum of n terms: $F_1 + F_2 + F_3 + \cdots + F_n = F_{n+2} - 1$
b Sum of n even terms: $F_2 + F_4 + F_6 + \cdots + F_{2n} = F_{2n+1} - 1$
c Sum of n odd terms: $F_1 + F_3 + F_5 + \cdots + F_{2n-1} = F_{2n}$
d Sum of the squares of the terms: $F_1^2 + F_2^2 + F_3^2 + \cdots + F_n^2 = F_n \times F_{n+1}$
e $F_n^2 = F_{n-1} \times F_{n+1} - 1$
f $F_n^2 + F_{n+1}^2 = F_{2n+1}$

11 **a** Prove that $\frac{d}{dx}(x^n) = nx^{n-1}, \quad \forall n \in N$.
[Hint: use the product rule.]
b Why, in this context, would proof by induction not be appropriate to show:
(i) $\frac{d}{dx}(x^n) = nx^{n-1}$, $\forall n \in Z$ **(ii)** $\frac{d}{dx}(x^n) = nx^{n-1}$, $\forall n \in Q$

12 Prove that, for all n greater than a particular value, $n! > 2^n$.
The particular value has to be stated in the first part of the proof.

13 Prove by induction that, with unlimited supplies of 4p and 7p stamps, you can make postage up to any value of 17p or more.

14 The fundamental theorem of arithmetic states that every natural number greater than 1 is either prime or can be expressed uniquely as the product of primes (see earlier). Prove by induction the lesser result that every natural number greater than 1 is either prime or can be expressed as the product of primes.
[Hint: the assumption to make is that every number up to $n = k$ is either prime or can be expressed as a product of primes.]

The binomial theorem revisited

In Advanced Higher Mathematics 1 the binomial theorem was introduced without proof. We are now able to prove this important theorem, at least for positive integers, by using proof by induction.

The binomial theorem states that

$$(a+b)^n = \binom{n}{0}a^n + \binom{n}{1}a^{n-1}b + \binom{n}{2}a^{n-2}b^2 + \dots + \binom{n}{r}a^{n-r}b^r + \dots + \binom{n}{n}b^n$$

Proof

1. The theorem is true when $n = 1$: $(a+b)^1 = \binom{1}{0}a^1 + \binom{1}{1}b^1 = a + b$.
2. Assume the theorem is true for $n = k$, where $k \in N$.

$$(a+b)^k = \binom{k}{0}a^k + \binom{k}{1}a^{k-1}b + \binom{k}{2}a^{k-2}b^2 + \dots + \binom{k}{r}a^{k-r}b^r + \dots + \binom{k}{k}b^k$$

The goal is

$$(a+b)^{k+1} = \binom{k+1}{0}a^{k+1} + \binom{k+1}{1}a^{k+1-1}b + \binom{k+1}{2}a^{k+1-2}b^2 + \dots$$
$$+ \binom{k+1}{r}a^{k+1-r}b^r + \dots + \binom{k+1}{k+1}b^{k+1}$$

Now $(a+b)^{k+1} = (a+b)(a+b)^k$

$$\Rightarrow (a+b)^{k+1} = (a+b)\left(\binom{k}{0}a^k + \binom{k}{1}a^{k-1}b + \binom{k}{2}a^{k-2}b^2 + \dots + \binom{k}{r}a^{k-r}b^r + \dots + \binom{k}{k}b^k\right)$$

using the assumption

$$= \binom{k}{0}a^{k+1} + \binom{k}{1}a^k b + \binom{k}{2}a^{k-1}b^2 + \dots + \binom{k}{r}a^{k-r+1}b^r + \dots + \binom{k}{k}ab^k$$
$$+ \binom{k}{0}a^k b + \binom{k}{1}a^{k-1}b^2 + \binom{k}{2}a^{k-2}b^2 + \dots + \binom{k}{r-1}a^{k-r+1}b^r + \dots + \binom{k}{k}b^{k+1}$$

$$= \binom{k}{0}a^{k+1} + \left(\binom{k}{1} + \binom{k}{0}\right)a^k b + \left(\binom{k}{2} + \binom{k}{1}\right)a^{k-1}b^2 + \dots$$
$$+ \left(\binom{k}{r} + \binom{k}{r-1}\right)a^{k-r+1}b^r + \dots + \binom{k}{k}b^{k+1}$$

$$\Rightarrow (a+b)^{k+1} = \binom{k+1}{0}a^{k+1} + \binom{k+1}{1}a^{k+1-1}b + \binom{k+1}{2}a^{k+1-2}b^2 + \dots$$
$$+ \binom{k+1}{r}a^{k+1-r}b^r + \dots + \binom{k+1}{k+1}b^{k+1}$$

Using the result from Book 1:
$\binom{n}{p} + \binom{n}{p-1} = \binom{n+1}{p}$

Thus if the conjecture is true for $n = k$ then it is true for $n = k + 1$.

Since it is true for $n = 1$, by induction it is true $\forall n \geq 1,\ n \in N$.

Remember that expressed using sigma notation the theorem reads:

$$(a + b)^n = \sum_{r=0}^{n} \binom{n}{r} a^{n-r} b^r$$

A short summary so far

You have seen the three types of proof: direct, indirect and induction.

- Direct proofs are sometimes the easiest to write down but often the most difficult to start.
- The indirect proofs explored were of two types:
  (i) proof by contradiction, where the negation is assumed and a contradiction is arrived at, establishing the truth of the original statement;
  (ii) proof by contrapositive, where the contrapositive is proved, thus establishing the truth of the original statement (which is logically equivalent to its contrapositive).
- The proofs by induction examined require a conjecture about whole numbers n. The mechanism requires that the conjecture is proved true for one value of n, $n = a$. One must then establish that if it is true for $n = k$ then it is true for $n = k + 1$. Tying these points together establishes the truth of the conjecture for all $n \geq a$.
- In any situation where a universal statement is made, it can be proved false by finding a counterexample.

  In any situation where an existential statement is made, it can be proved true by finding an example.

Harder mixed questions

EXERCISE 5

1 **a** Form a direct proof that 9^n can be expressed in the form $8k + 1$ where $n, k \in N$.
b Hence prove by contradiction that $3^{2n} + 5$ is never divisible by 8.

2 The arithmetic/geometric mean (A/GM) inequality with three terms looks like this

$$\frac{a_1 + a_2 + a_3}{3} \geq \sqrt[3]{a_1 a_2 a_3}$$

where $\frac{a_1 + a_2 + a_3}{3}$ is the *arithmetic mean* and $\sqrt[3]{a_1 a_2 a_3}$ is the *geometric mean* of the three numbers a_1, a_2, a_3.
You are given that p, q and r are three non-negative real numbers such that $(1 + p)(1 + q)(1 + r) = 8$.
a Show that the given statement is equivalent to $1 + (p + q + r) + (pq + qr + rp) + pqr = 8$.
b Use the A/GM inequality with $a_1 = p$, $a_2 = q$, $a_3 = r$ to form a useful inequality.
c Use the A/GM inequality with $a_1 = pq$, $a_2 = qr$, $a_3 = rp$ to form another useful inequality.
d Combine the above three parts to form a direct proof that $pqr < 1$.

3 Prove that $(a + b)^3 \geq \frac{27}{4}a^2b,\ \forall a, b \geq 0$.
[Hint: try a couple of values for a and b just to convince yourself that the conjecture is true and get a feel for the inequality. Use the A/GM inequality for three terms. How can you usefully express a and b as three terms? Find this and the proof takes only two or three lines.]

4 Prove by contradiction that, in every tetrahedron, there is a vertex such that the three edges meeting at the vertex have lengths which could be used to construct a triangle.

> *Hints*
> Nominate a largest side: let AB be the largest side.
> Make the assumption 'there is no vertex such that …'.
> Use the triangle inequality, namely that if a, b, c are the sides of a triangle then $a < b + c$, and the inverse: if a, b, c cannot form a triangle then $a \geq b + c$. At vertex A, AB ≥ AC + AD and at vertex B …
> Now use the triangle inequality on triangles which have AB as a side.

5 Fermat's little theorem states that, if x and p have no common factor, and p is prime, then $x^p - 1$ is divisible by p. Use this theorem to give a direct proof of the following:

Every square number S has the form $5n$ or $5n + 1$ or $5n - 1$, $n \in N$.

> *Hint*
> Start:
> Either S is divisible by 5 or it isn't.
> So either $S = 5k$ or S and 5 have no common factor …

CHAPTER 1 REVIEW

1 Form a direct proof that $2 + 4 + 6 + \cdots + 2n = n(n + 1)$ where $n \in W$.

2 Prove directly that $[(a + b)^2 - (a - b)^2 > 0 \quad \textit{and} \quad a > 0] \Rightarrow b > 0$.

3 Give a counterexample to prove that the following statement is false:
'If the diagonals of a quadrilateral intersect at 90° then the quadrilateral is a rhombus.'

4 Prove by contradiction that $\sqrt{11}$ is irrational.

5 Prove by induction that $2 + 5 + 8 + \cdots + (3n - 1) = \frac{1}{2}n(3n + 1)$, for all $n \geq 1$, $n \in N$.

6 **a** Prove by induction that $n(n^2 - 1)(3n + 2)$ is divisible by 24.

b **(i)** Prove directly that the product of two consecutive numbers is divisible by 2.

(ii) Prove directly that the product of three consecutive numbers is divisible by 6.

(iii) Hence prove directly that $n(n^2 - 1)(3n + 2)$ is divisible by 24.

CHAPTER 1 SUMMARY

1 **(i)** Let a, b, c represent *statements* which can be either true or false but not both.

(ii) The *negation* of a, *not-a*, is true when a is false and false when a is true.

(iii) Statements can be combined to form compound statements:

a and b, a or b, If a then b

(iv) A *universal* statement refers to all items in a set: 'for all x' is written $\forall x$.
A universal statement can be disproved if one *counterexample* can be cited.

(v) An *existential* statement refers to the existence of an item in a set:
'there exists an x' is written $\exists x$.
An existential statement can be proved if one example can be cited.

2 **(i)** 'If a then b' is an *implication* ($a \Rightarrow b$).

(ii) 'If *not-a* then *not-b*' is the *inverse* of 'If a then b'.

(iii) 'If b then a' is the *converse* of 'If a then b' ($b \Rightarrow a$).

(iv) 'If *not-b* then *not-a*' is the *contrapositive* of 'If a then b'.

(v) If both an implication and its converse are true the statements are said to be *equivalent* ($a \Leftrightarrow b$).

(vi) An implication and its contrapositive are logically equivalent.

3 **(i)** In *direct proof* we move directly from a statement whose truth is given, by a series of implications, to the truth of another statement. For example

Given a is true, $a \Rightarrow b$, $b \Rightarrow c$, $c \Rightarrow d$ proves d is true.

(ii) *Indirect proof – proof by contradiction*
The negation of that which we wish to prove is assumed, a contradiction is reached, which implies the assumption is false.

(iii) *Indirect proof – proof by contrapositive*
Rather than prove $a \Rightarrow b$ it can be easier to prove *not-b* $\Rightarrow$ *not-a*, which is equivalent to it.

(iv) *Proof by induction*
Wishing to prove some statement involving natural numbers n:

(a) we demonstrate its truth when $n = a$, some particular value of n;

(b) we prove that 'true for $n = k$' implies 'true for $n = k + 1$';

(c) **(a)** and **(b)** combine to prove by induction that the statement is true $\forall n \geq a$, $n \in N$.

2.1 Further Differentiation

Historical note

Euler

Leonhard Euler (1707–1783) was a Swiss mathematician but for most of his career he worked in St Petersburg in Russia. Much of the algebraic notation we use today is his invention: π, e, Σ, $f(x)$, etc.

Euler's first book *Introductio Analysin Infinitorum* was published in 1748. Part one of the book deals with algebraic analysis and in it Euler introduces the concept of a function as we know it and makes the distinction between explicit and implicit functions.

Part two of the book deals with coordinate geometry and in this section Euler introduces the idea of parametric equations.

Reminders

The chain rule

$\dfrac{dy}{dx} = \dfrac{dy}{du}\dfrac{du}{dx}$ or $\dfrac{d}{dx}f(g(x)) = f'(g(x)).g'(x)$

The inverse function

If f has an inverse then it is denoted by f^{-1}, and $f(f^{-1}(x)) = x = f^{-1}(f(x))$ for all x in the domain of f.

If $y = f(x)$ is the equation of a function then $x = f(y)$ provides the equation of the inverse function.

The graph of the inverse function is the image of the graph of the function under reflection in the line $y = x$.

The inverse trigonometric functions

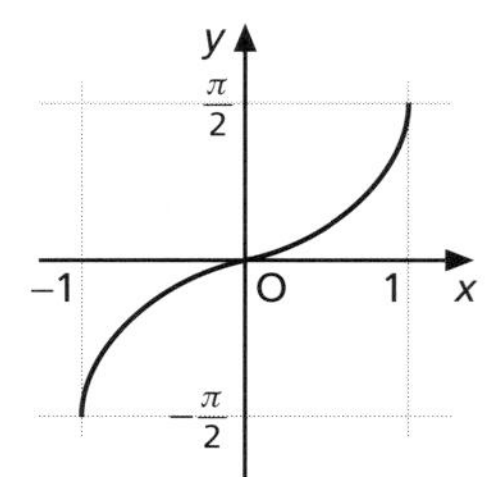

The inverse sine
$y = \sin^{-1} x$
Domain [−1, 1]
Sometimes called the arcsine and accessed on a spreadsheet by ASIN()

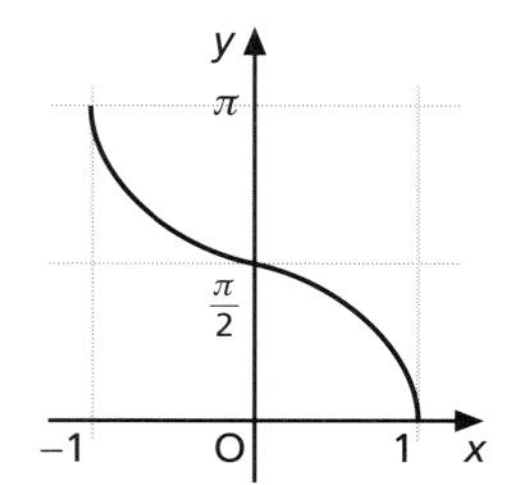

The inverse cosine
$y = \cos^{-1} x$
Domain [−1, 1]
Sometimes called the arccosine and accessed on a spreadsheet by ACOS()

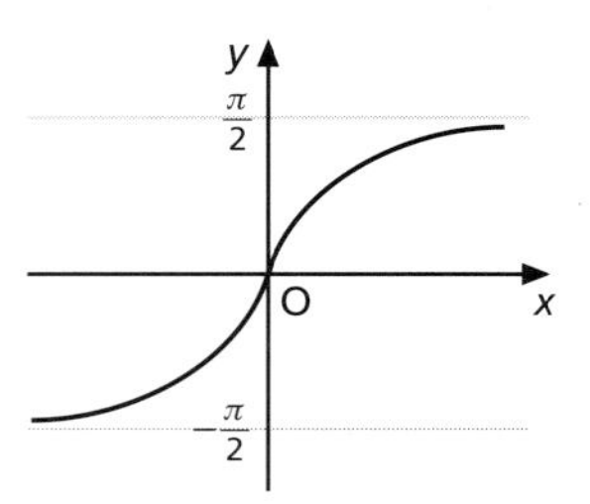

The inverse tangent
$y = \tan^{-1} x$
Domain $(-\infty, \infty)$
Sometimes called the arctangent and accessed on a spreadsheet by ATAN()

Derivative of inverse functions

Consider the function f which has an inverse f^{-1}. Then

$$f(f^{-1}(x)) = x$$

Differentiating using the chain rule:

$$f'(f^{-1}(x)).\frac{d}{dx}f^{-1}(x) = 1$$

thus $$\frac{d}{dx}f^{-1}(x) = \frac{1}{f'(f^{-1}(x))}$$

Example 1 Given $f(x) = x^3$ find $f'(x)$ and state the derivative of $f^{-1}(x)$.

$$f'(x) = 3x^2$$

$$f^{-1}(x) = x^{\frac{1}{3}}$$

$$\Rightarrow \frac{d}{dx}f^{-1}(x) = \frac{1}{f'(f^{-1}(x))} = \frac{1}{3\left(x^{\frac{1}{3}}\right)^2} = \frac{1}{3x^{\frac{2}{3}}}$$

This result can be more easily obtained by differentiating f^{-1} directly. It is put forward here to illustrate a technique.

Example 2 Express $f(x) = x^2 + 2x + 3,\ x \geq 3$, in the form $p(x + q)^2 + r$.
Find $f'(x)$ and state the derivative of $f^{-1}(x)$.

Completing the square we get $f(x) = (x + 1)^2 + 2$.

Hence $f^{-1}(x) = \sqrt{(x - 2)} - 1$, taking the positive root since $x \geq 3$ is the domain of $f(x)$.

$$f'(x) = 2x + 2$$

$$\frac{d}{dx}f^{-1}(x) = \frac{1}{2(\sqrt{(x-2)} - 1) + 2} = \frac{1}{2\sqrt{(x-2)}}$$

Example 3 Given $f(x) = \sin x,\ -\frac{\pi}{2} \leq x \leq \frac{\pi}{2}$, find the derivative of $f^{-1}(x)$.

$$f'(x) = \cos x \Rightarrow \frac{d}{dx}\sin^{-1} x = \frac{1}{\cos(\sin^{-1}(x))}$$

If $\sin^{-1} x = a$ then $x = \sin a$ and since $|a| \leq \frac{\pi}{2}$ there exists a triangle with hypotenuse 1, and sides x and $\sqrt{(1 - x^2)}$ (by Pythagoras) as shown.
Thus $\cos(\sin^{-1} x) = \cos a = \sqrt{(1 - x^2)}$.

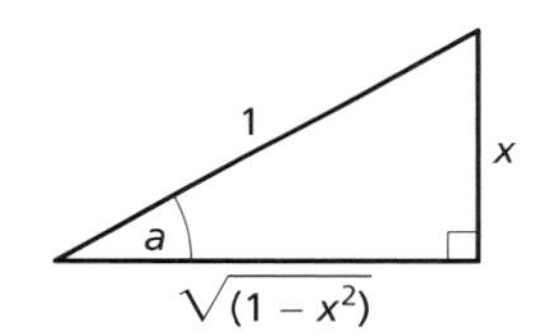

$$\boxed{\frac{d}{dx}\sin^{-1} x = \frac{1}{\sqrt{(1 - x^2)}}}$$

EXERCISE 1A

1 For each function $f(x)$, find the derivative of the inverse function $f^{-1}(x)$ using the technique illustrated in Example 1.

a $f(x) = x^5$ **b** $f(x) = x^{\frac{3}{4}}$ **c** $f(x) = 2x^{-2},\ x > 0$ **d** $f(x) = x^2 + 1,\ x > 1$

2 In each case below
(i) express $f(x)$ in the form $p(x + q)^2 + r$;
(ii) find $f^{-1}(x)$, taking the positive root;
(iii) find $f'(x)$ and state the derivative of $f^{-1}(x)$.

a $f(x) = x^2 + 4x + 5,\ x \geq 5$ **b** $f(x) = x^2 + 6x - 1,\ x \geq -1$
c $f(x) = 2x^2 - 2x + 1,\ x \geq 1$ **d** $f(x) = 3x^2 + 2x,\ x \geq 0$

3 a By considering $f(x) = e^x$, deduce that the derivative of $\ln x,\ x > 0$, is x^{-1}.
b If $g(x) = e^{2x+1}$ find the derivative of $\frac{1}{2}(\ln x - 1)$ by this technique.

4 Use the techniques shown in Example 3 to find the derivative of
a $\cos^{-1} x$
b $\tan^{-1} x$ $\left[\text{Remember that } \frac{d}{dx}\tan x = \sec^2 x.\right]$

A useful result

Given a function f and its inverse f^{-1} defined over a suitable interval, we have

$$y = f(x) \Rightarrow x = f^{-1}(y)$$

$$\Rightarrow \frac{dx}{dy} = \frac{1}{f'(f^{-1}(y))} = \frac{1}{f'(x)} = \frac{1}{\frac{dy}{dx}}$$

$$\boxed{\frac{dx}{dy} = \frac{1}{\frac{dy}{dx}}}$$

Example 1 Given $y = \cos^{-1} x$, $-1 \le x \le 1$, find $\frac{dy}{dx}$.

$$y = \cos^{-1} x \Rightarrow x = \cos y,\ 0 \le y \le \pi$$

$$\frac{dx}{dy} = -\sin y$$

$$= -\sqrt{(1 - \cos^2 y)} = -\sqrt{(1 - x^2)}$$

$$\frac{dy}{dx} = \frac{-1}{\sqrt{(1 - x^2)}}$$

Note

$\sin y = \pm\sqrt{(1 - \cos^2 y)}$. Since $\cos^{-1} x$ is defined in the range $0 \le y \le \pi$, and throughout this range the gradient of $y = \cos^{-1} x$ is negative, the positive root is taken.

Example 2 Given $y = \ln x$, find $\frac{dy}{dx}$.

$$y = \ln x \Rightarrow x = e^y$$

$$\frac{dx}{dy} = e^y = x$$

$$\Rightarrow \frac{dy}{dx} = \frac{1}{x}$$

EXERCISE 1B

1 Use the above technique to find $\frac{dy}{dx}$ for each of the following.

a $\tan^{-1} x$

[You will need to know that $\sec^2 x = 1 + \tan^2 x$.]

b $\sin^{-1}(3x + 1)$, $-\frac{1}{3} \le x \le 0$

2 For each of the following

(i) express y in the form $p(x+q)^2+r$;

(ii) make x the subject of the formula;

(iii) find $\dfrac{dx}{dy}$;

(iv) find $\dfrac{dy}{dx}$ and verify that $\dfrac{dx}{dy}=\dfrac{1}{\dfrac{dy}{dx}}$.

a $y=x^2+6x+1,\ x\geq 1$

b $y=x^2+2x-5,\ x\geq -5$

c $y=3x^2-4x-3,\ x\geq -3$

d $y=2x^2+4x,\ x\geq 0$

3 When using spreadsheets or graphics calculators it is possible to get a good approximation for the graph of the derivative of a function.
If $f(x)$ is a function then

$$f'(x)\approx\frac{(f(x+0.0005)-f(x-0.0005))}{0.001}$$

a Compare the graph $y=\dfrac{(\sin^{-1}(x+0.0005)-\sin^{-1}(x-0.0005))}{0.001}$ with $y=\dfrac{1}{\sqrt{(1-x^2)}}$.

b Consider the derivatives of **(i)** $\cos^{-1}(x)$, **(ii)** $\tan^{-1}(x)$, **(iii)** $\ln x$ in a similar way.

Inverse trigonometric functions

Applying the chain rule to inverse trigonometric functions

$$\frac{d}{dx}\sin^{-1}x=\frac{1}{\sqrt{(1-x^2)}}\qquad\frac{d}{dx}\cos^{-1}x=\frac{-1}{\sqrt{(1-x^2)}}\qquad\frac{d}{dx}\tan^{-1}x=\frac{1}{1+x^2}$$

Many examples using these three standard derivatives also involve the chain rule.

Example 1 Differentiate $\sin^{-1}\left(\dfrac{-1}{x}\right)$.

$$\frac{d}{dx}\sin^{-1}\left(\frac{-1}{x}\right)=\frac{1}{\sqrt{\left[1-\left(\frac{-1}{x}\right)^2\right]}}\frac{d}{dx}\left(\frac{-1}{x}\right)=\frac{1}{\sqrt{\frac{x^2-1}{x^2}}}\frac{1}{x^2}=\frac{1}{x\sqrt{(x^2-1)}}$$

Example 2 Differentiate $\ln(1+\cos^{-1}x)$.

$$\frac{d}{dx}\ln(1+\cos^{-1}x)=\frac{1}{1+\cos^{-1}x}\frac{d}{dx}(1+\cos^{-1}x)=\frac{1}{1+\cos^{-1}x}\frac{-1}{\sqrt{(1-x^2)}}$$

$$=\frac{-1}{(1+\cos^{-1}x)\sqrt{(1-x^2)}}$$

EXERCISE 2

1 Find the derivative of:

a $\sin^{-1} x^2$ **b** $\tan^{-1}(x+2)$ **c** $\sin^{-1}\frac{1}{x}$ **d** $\tan^{-1}\frac{1}{\sqrt{x}}$

e $\cos^{-1}\frac{1}{x}$ **f** $\cos^{-1} ax$

2 Find the derived function of:

a $\sin^{-1}(e^x)$ **b** $\cos^{-1}(x+2)^2$ **c** $\sin^{-1}\sqrt{(1-x^2)}$ **d** $\sin^{-1}(\tan x)$

e $\sin^{-1}\left(\frac{x}{a}\right)$

3 Find $f'(x)$ for each of the following expressions for $f(x)$.

a $\cos^{-1} e^{2x}$ **b** $\sin^{-1}\cos(x-1)$ **c** $\tan^{-1}(1+x)$ **d** $\cos^{-1}(\ln 3x)$

e $\sec^{-1} 3x$

4 Differentiate:

a $\ln(\tan^{-1}\sqrt{x})$ **b** $\ln\left(\sin^{-1}\frac{-1}{\sqrt{x}}\right)$ **c** $\ln\sin^{-1} e^x$ **d** $e^{\sin^{-1} x}$

5 Calculate:

a $f'(1)$ where $f(x) = e^{\tan^{-1}\left(\frac{1}{x}\right)}$ **b** $f'(0)$ where $f(x) = \ln\cos^{-1} x$

c $f'\left(\frac{3}{4}\right)$ where $f(x) = \sin(\tan^{-1} x)$ **d** $f'(\sqrt{3})$ where $f(x) = \cos\left(\tan^{-1}\frac{1}{x}\right)$

6 Find the gradient of the curve with equation:

a $y = \ln(\sin^{-1} 2x)$ where $x = \frac{1}{4}$ **b** $y = [x + \tan^{-1} x]^3$ where $x = 1$

c $y = \ln(\cos^{-1}(1-x))$ where $x = \frac{1}{2}$ **d** $y = e^{\tan^{-1} x^2}$ where $x = 1$

7 **a** For what values of x does $y = \ln\frac{2}{\tan^{-1} x}$ exist?

b Show that in this domain it is a decreasing function.

8 Show that $y = [\sin^{-1} 3x]^4$ has a minimum turning point at the origin.

9 **a** For what range of values of x does the function given by $f(x) = [1 + \cos^{-1} 2x]^{\frac{1}{2}}$ exist?

b Show that this function is decreasing throughout its domain.

Applying the product and quotient rules to inverse trigonometric functions

Example 1 Differentiate $x^2\cos^{-1}\sqrt{x}$.

$$\frac{d}{dx}(x^2\cos^{-1}\sqrt{x}) = x^2\frac{d}{dx}(\cos^{-1}\sqrt{x}) + \frac{d}{dx}(x^2)\cos^{-1}\sqrt{x} \quad \text{by the product rule}$$

$$= x^2\frac{-1}{\sqrt{(1-x)}}\,\frac{1}{2}x^{-\frac{1}{2}} + 2x\cos^{-1}\sqrt{x} \quad \text{by the chain rule}$$

$$= 2x\cos^{-1}\sqrt{x} - \frac{x^{\frac{3}{2}}}{2\sqrt{(1-x)}}$$

> *Example 2* Find the derivative of $\tan^{-1}\left(\dfrac{1+\sin x}{1-\cos x}\right)$.
>
> $$\frac{\mathrm{d}}{\mathrm{d}x}\tan^{-1}\left(\frac{1+\sin x}{1-\cos x}\right)$$
>
> $$= \frac{1}{1+\left(\dfrac{1+\sin x}{1-\cos x}\right)^2}\frac{\mathrm{d}}{\mathrm{d}x}\left(\frac{1+\sin x}{1-\cos x}\right) \quad \text{by the chain rule}$$
>
> $$= \frac{1}{1+\left(\dfrac{1+\sin x}{1-\cos x}\right)^2}\,\frac{(1-\cos x)\cos x-(1+\sin x)\sin x}{(1-\cos x)^2} \quad \text{by the quotient rule}$$
>
> $$= \frac{\cos x-\cos^2 x-\sin x-\sin^2 x}{(1-\cos x)^2+(1+\sin x)^2}$$
>
> $$= \frac{\cos x-\sin x-1}{3+2\sin x-2\cos x} \quad \text{using } \sin^2 x+\cos^2 x=1$$

EXERCISE 3A

1 Differentiate:

a $x^2\sin^{-1}x$ **b** $x\sin^{-1}x^2$ **c** $\sqrt{x}\cos^{-1}x$ **d** $\sqrt{x}\sin^{-1}\sqrt{x}$

2 Find the derivative of:

a $(1+x^2)\tan^{-1}x$ **b** $e^x\sin^{-1}x$ **c** $e^{2x}\cos^{-1}\left(\dfrac{x}{2}\right)$ **d** $\ln x\tan^{-1}x$

3 Find the derived function $f'(x)$ for each of the following.

a $f(x)=\dfrac{\tan^{-1}x}{x}$ **b** $f(x)=\dfrac{\sin^{-1}x}{\sqrt{x}}$ **c** $f(x)=\dfrac{\cos^{-1}2x}{x\sqrt{x}}$

d $f(x)=\dfrac{\tan^{-1}(x+1)}{x^2}$

4 Find $f'(x)$ for each of the following.

a $f(x)=\dfrac{x}{\sin^{-1}x}$ **b** $f(x)=\dfrac{x^2}{\cos^{-1}(x-1)}$ **c** $f(x)=\dfrac{e^x}{\sin^{-1}2x}$

d $f(x)=\dfrac{\ln x}{\tan^{-1}x}$

5 Calculate the gradient of the tangent to the curve with equation:

a $y=\tan^{-1}\left(\dfrac{2x+3}{3x-2}\right)$ at $x=-\dfrac{1}{2}$ **b** $y=\sin^{-1}\left(\dfrac{1+2\cos x}{2+\cos x}\right)$ at $x=\dfrac{\pi}{6}$

6 Find the coordinates of the stationary point on the curve with equation:

a $y=\tan^{-1}\left(\dfrac{e^x}{x}\right)$ **b** $y=\cos^{-1}\left(\dfrac{\ln x}{x}\right)$

EXERCISE 3B

1 Show that there are no turning points on the curve with equation $y = \sin^{-1}\left(\frac{1-x}{1+x}\right)$.

2 Express the derivative of $\tan^{-1}\left(\frac{2\sin x + \sin 2x}{2\cos x + \cos 2x}\right)$ in terms of $\cos x$.

3 **a** Show that $f(x) = \cos^{-1}\left(\frac{1-x^2}{1+x^2}\right)$ and $g(x) = 2\tan^{-1} x$ have the same derived function.

b This could be because $f(x) = g(x)$. Prove, by trigonometry, that this is in fact true.

4 Find the gradient of the curve $y = \sin^{-1}\left(\frac{x^3-1}{x^3+1}\right)$ at $x = 0$ and show that the gradient approaches this same value as $x \to \infty$.

5 Show that

a $y = (x-1)\cos^{-1}\left(\frac{x-1}{x}\right) \Rightarrow x(x-1)\sqrt{(2x-1)}\frac{dy}{dx} = x\sqrt{(2x-1)}y - (x-1)^2$

b $y = (x+1)\sin^{-1}\left(\frac{x}{x+1}\right) \Rightarrow (x+1)\sqrt{(2x+1)}\frac{dy}{dx} = \sqrt{(2x+1)}y + (x+1)$

c $y = x^2\tan^{-1}\left(\frac{x-1}{x+1}\right) \Rightarrow 2x\frac{dy}{dx} = \frac{2x^3}{x^2+1} + 4y$

d $y = e^x\cos^{-1}\left(\frac{x-1}{x+1}\right) \Rightarrow 2\sqrt{x}(x+1)\left(\frac{dy}{dx} - y\right) + 2e^x = 0$

e $y = 2x\sin^{-1}\left(\frac{\cos x}{x}\right) \Rightarrow x\frac{dy}{dx} - y = \frac{-2x(x\sin x + \cos x)}{\sqrt{(x^2 - \cos^2 x)}}$

f $y = x\tan^{-1}\left(\frac{e^x}{x}\right) \Rightarrow x\frac{dy}{dx} - y = \frac{x^2(x-1)e^x}{x^2 + e^{2x}}$

Implicit and explicit functions

The equations $3x + 4y = 24$ and $y = -\frac{3}{4}x + 6$ are different forms of the equation of a line. In the latter form, y is expressed *explicitly* as a function of x.
In the form $3x + 4y = 24$, y is still a function of x, but this fact is expressed *implicitly.*

When the dependent variable, y, is expressed in terms of the independent variable, x (i.e. it is the subject of the formula), the function is said to be *explicit.* When the dependent variable is not the subject of the formula, then the function is said to be *implicit.*

The equation of the circle in the form $x^2 + y^2 = 25$ is another example of an implicit function. When y is made the subject, $y = \pm\sqrt{(25 - x^2)}$, giving two explicit functions of x.

In such cases the implicit function is referred to as a *multiple-valued* function.

First derivatives of implicit functions

When a function is stated implicitly, it is often possible to find the derived function without having to find the explicit form first.

Example 1 Find the gradient of the tangent at A(3, −4) to the circle with equation $x^2 + y^2 = 25$.

Differentiating each term with respect to x,

$$2x + 2y\frac{dy}{dx} = 0$$

$$\Rightarrow \quad \frac{dy}{dx} = -\frac{x}{y}$$

$$\Rightarrow \text{ the gradient of the tangent at A}(3, -4) = \frac{-3}{-4} = \frac{3}{4}$$

Since y is a function of x, the chain rule must be applied to obtain the derivative of y^2

This result can be verified by purely geometrical considerations.

Example 2 Find the gradient of the tangent at any point on the lemniscate (the figure-8 curve) with equation $x^4 = x^2 - y^2$.

Differentiating gives

$$4x^3 = 2x - 2y\frac{dy}{dx}$$

$$\Rightarrow \quad \frac{dy}{dx} = \frac{x(1 - 2x^2)}{y}$$

It is of interest to obtain the explicit form of this equation and sketch the curve.

$$y^2 = x^2 - x^4 = x^2(1 - x^2)$$

$$\Rightarrow \quad y = \pm x\sqrt{(1 - x^2)}$$

Applying your curve-sketching skills or using a graphic calculator reveals this graph.

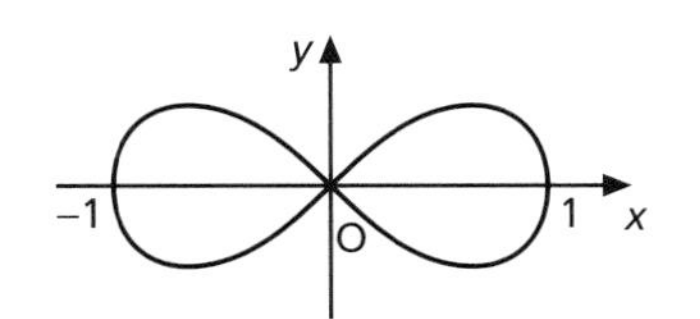

Example 3 For the curve with equation $x^3 - xy + y^2 = 1$, express $\frac{dy}{dx}$ in terms of x and y.

Note that the product rule is required when differentiating the term xy.

$$3x^2 - \left(x\frac{dy}{dx} + 1.y\right) + 2y\frac{dy}{dx} = 0$$

$$\Rightarrow \quad 3x^2 - y + (2y - x)\frac{dy}{dx} = 0$$

$$\Rightarrow \quad \frac{dy}{dx} = \frac{y - 3x^2}{2y - x}$$

This function can also be expressed explicitly by re-arranging the equation as $y^2 - xy + (x^3 - 1) = 0$ and applying the quadratic formula to give $y = \frac{1}{2}\left[x \pm \sqrt{(4 + x^2 - 4x^3)}\right]$.

A graphic calculator will exhibit two branches meeting at the point (1, 1).

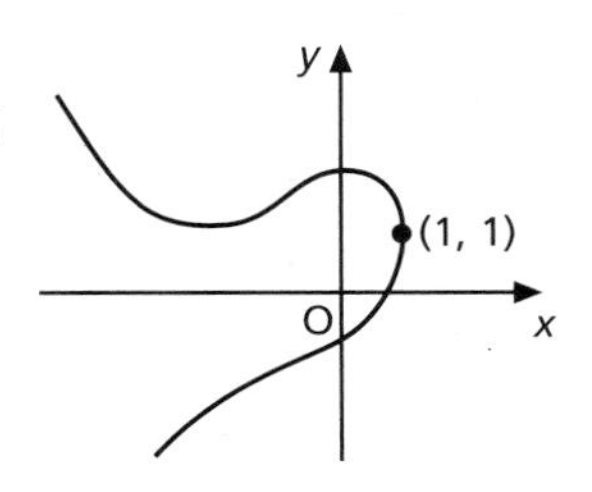

Example 4 If $(x + y)^3 = 3xy$, show that $\dfrac{dy}{dx} = \dfrac{y(y - 2x)}{x(2y - x)}$.

$$3(x + y)^2\left(1 + \frac{dy}{dx}\right) = 3\left(x\frac{dy}{dx} + 1.y\right)$$ using the chain and product rules

$$\Rightarrow \quad (x + y)^2 + (x + y)^2\frac{dy}{dx} = x\frac{dy}{dx} + y$$

$$\Rightarrow \quad \frac{dy}{dx} = \frac{y - (x + y)^2}{(x + y)^2 - x}$$

$$= \frac{(x + y)y - (x + y)^3}{(x + y)^3 - (x + y)x}$$ multiplying numerator and denominator by $(x + y)$

$$= \frac{xy + y^2 - 3xy}{3xy - x^2 - xy}$$ using the initial equation

$$= \frac{y(y - 2x)}{x(2y - x)}$$

EXERCISE 4A

1 Find $\dfrac{dy}{dx}$ for each of the following implicit functions of x.

a $x^2 + 4xy + y^2 = 8$
b $x^2 = \ln y$
c $2x^2 + 2y^2 - 5x + 4y - 9 = 0$
d $x + 3 = e^y$
e $x^2 - xy + 3y^2 = 10$
f $x \tan y = e^x$
g $x^{\frac{2}{5}} + y^{\frac{2}{5}} = e^{\frac{2}{5}}$
h $\ln(x + y) = \tan^{-1} x$
i $5x^2 - 4xy + 3y^2 = 2$
j $\sin^{-1} x + \cos^{-1} y = 2x^3$

2 a (i) Given $x = \sin y$, find $\dfrac{dy}{dx}$ in terms of y.
(ii) Hence express $\dfrac{dy}{dx}$ in terms of x.
b Similarly deduce the derivative of: **(i)** $y = \cos^{-1} x$ **(ii)** $y = \tan^{-1} x$.

3 Find an expression in terms of x and y for the gradient at the point (x, y) on the curve with equation $x^2 + y^2 = \dfrac{y}{x}$.

4 Prove that the curve $e^x + e^y - e = \dfrac{x}{y} + 1$ has a tangent at the point $(0, 1)$ which is parallel to the x-axis.

5 Find the equation of the tangent to the curve with equation $xy^4 + 3x^2y^2 = 28$ at the point $(1, 2)$.

6 Show that there is no point on the curve with equation $x + y = \ln(x - y)$ where the tangent is at 45° to the x-axis.

7 For the curve with equation $x \ln y = \cos x + \cos y$, show that the gradient at $x = 0$ is not defined.

8 Show that $x^2 = y^2 \ln y \Rightarrow \dfrac{dy}{dx} = \dfrac{2xy}{y^2 + 2x^2}$.

9 Find the equation of the tangent to the curve with equation $(x + 2y)^3 - 4x - 3y = 5$ at the point where it crosses the y-axis.

Harder questions

EXERCISE 4B

1 Find the gradient of the tangent at the point (e, e^2) on the curve given by $x \ln x + y \ln y = e(1 + 2e)$.

2 Show that $y = x - 1$ is the equation of the tangent at the point $\left(\frac{1}{2}, -\frac{1}{2}\right)$ to the curve with equation $\sin^{-1} x + \cos^{-1} y = \dfrac{5\pi}{6}$.

3 **a** Find the points of intersection of the parabola with equation $y^2 = 2x$ and the circle with equation $x^2 + y^2 = 8$.

b Use implicit differentiation to find the gradients of the tangents at the point of intersection in the first quadrant, and calculate the size of the angle between them.

4 Show that the point A(3, 2) lies on the circle with equation $(x - 6)^2 + (y - 5)^2 = 18$ and the parabola with equation $y^2 = 16 - 4x$ and that these curves share a common tangent at A.

5 **a** Find the coordinates of the four points of intersection of the ellipse with equation $\dfrac{x^2}{25} + \dfrac{y^2}{4} = 25$ and the hyperbola with equation $\dfrac{x^2}{16} - \dfrac{y^2}{9} = 21$.

b Show that these curves are *orthogonal,* i.e. at their points of intersection, their tangents are perpendicular.

6 Show that, for $e^{x+y} = \ln(x - y)$, $\dfrac{dy}{dx}$ can be written as $\dfrac{1 - e^{(a+x+y)}}{1 + e^{(a+x+y)}}$ and state the appropriate expression for a.

7 Show that, if $x + y = \dfrac{x}{y}$, then $\dfrac{dy}{dx}$ can be written as $\dfrac{y^3}{x^2(2 - y)}$.

Second derivatives of implicit functions

Higher-order derivatives can also be found implicitly.

Example 1 Find $\dfrac{dy}{dx}$ and $\dfrac{d^2y}{dx^2}$ in terms of x and y only, for the function $y(x)$ defined implicitly by $x^2 + 2xy = 1$.

Differentiating

$$2x + 2\left(x\frac{dy}{dx} + 1.y\right) = 0$$

$$\Rightarrow \quad x + y + x\frac{dy}{dx} = 0$$

$$\Rightarrow \quad \frac{dy}{dx} = -\frac{x + y}{x}$$

Differentiating $x + y + x\dfrac{dy}{dx} = 0$ gives

$$1 + \frac{dy}{dx} + \left[x\frac{d^2y}{dx^2} + 1.\frac{dy}{dx}\right] = 0$$

$$\Rightarrow \quad 1 + 2\frac{dy}{dx} + x\frac{d^2y}{dx^2} = 0$$

$$\Rightarrow \quad x\frac{d^2y}{dx^2} = -2\frac{dy}{dx} - 1$$

$$= \frac{2(x+y)}{x} - 1 = \frac{2x + 2y - x}{x}$$

$$\Rightarrow \quad \frac{d^2y}{dx^2} = \frac{x+2y}{x^2}$$

EXERCISE 5

1 Find $\dfrac{dy}{dx}$ and $\dfrac{d^2y}{dx^2}$ in terms of x and y only, for each of these implicit functions.

a $x^2 + xy = 3$
b $x^3 + y^2 = 5$
c $\sqrt{x} + y^3 = 1$
d $xy = y^2 + 2$
e $(x+1)(y-1) = e^x$
f $xy^2 + 2 = y$
g $\ln(x+y) = x - y$
h $x^2 = y \ln y$
i $(x+y)^2 = e^y$
j $y = x(y + \sin x)$
k $y = \sin(x+y)$

2 Find the gradient of the tangent at the point A(1, e) on the curve defined by $x \ln y + y \ln x = 1$ and determine whether the curve is concave up or concave down at A. (Remember: if $f''(a) > 0$ then the curve is concave up at $x = a$.)

3 Find the gradient of the tangent at the point $B\left(\dfrac{\pi}{2}, \dfrac{\pi}{2}\right)$ on the curve defined by $x \cos y + y \cos x = 0$ and determine whether the curve is concave up or concave down at B.

4 Show that, if $x^2 = e^y$, then $e^y\dfrac{d^2y}{dx^2} + 2x\dfrac{dy}{dx} = 2$.

5 If $y = xe^y$, show that $(1-y)\dfrac{d^2y}{dx^2} = (2-y)\left(\dfrac{dy}{dx}\right)^2$.

6 For the function defined implicitly by $x^3 - xy + y^2 = 1$ evaluate $\dfrac{dy}{dx}$ at (1, 1) and $\dfrac{d^2y}{dx^2}$ at (1, 0).

7 For the function $y(x)$ defined implicitly by $y \cos x = e^x$ evaluate $y'\left(\dfrac{\pi}{3}\right)$ and $y''\left(\dfrac{\pi}{3}\right)$.

8 Find two points on the curve with equation $x^2 - xy + y^2 = 3$ where the tangent is parallel to the x-axis, and two points where the tangent is parallel to the y-axis.

9 Find two stationary points on the curve with equation $2x^2 - xy + 3y^2 = 46$, and by using the second derivative determine the nature of each.

Logarithmic differentiation

When a function is complicated by the occurrence of powers, roots, products and quotients of several factors it is useful to take logarithms before differentiating.

Reminders

$\ln(x \times y) = \ln x + \ln y$

$\ln(x \div y) = \ln x - \ln y$

$\ln x^r = r \ln x$

Example 1 Differentiate 2^{x+3}.

Let $y = 2^{x+3}$.

$$\ln y = \ln 2^{x+3}$$ taking the logarithms of both sides

$$\Rightarrow \ln y = (x+3)\ln 2$$

$$\Rightarrow \frac{1}{y}\frac{dy}{dx} = 1 \times \ln 2$$

$$\Rightarrow \frac{dy}{dx} = y \ln 2$$

$$\Rightarrow \frac{dy}{dx} = 2^{x+3}\ln 2$$

Example 2 Find $\frac{dy}{dx}$ when $y = x^x$.

$$y = x^x \Rightarrow \ln y = \ln x^x = x \ln x$$

$$\Rightarrow \frac{1}{y}\frac{dy}{dx} = x \times \frac{1}{x} + 1 \times \ln x = 1 + \ln x$$

$$\Rightarrow \frac{dy}{dx} = (1 + \ln x)y$$

$$\Rightarrow \frac{dy}{dx} = (1 + \ln x)x^x$$

Example 3 Find $\frac{dy}{dx}$ when $y = \frac{(2x+1)^{\frac{1}{2}}(3x-1)^{\frac{2}{3}}}{(4x+3)^{\frac{3}{4}}}$.

$$\ln y = \tfrac{1}{2}\ln(2x+1) + \tfrac{2}{3}\ln(3x-1) - \tfrac{3}{4}\ln(4x+3)$$ using the laws of logs

$$\Rightarrow \frac{1}{y}\frac{dy}{dx} = \frac{1}{2}\,\frac{2}{2x+1} + \frac{2}{3}\,\frac{3}{3x-1} - \frac{3}{4}\,\frac{4}{4x+3}$$

$$= \frac{1}{2x+1} + \frac{2}{3x-1} - \frac{3}{4x+3}$$ tidying up

$$\Rightarrow \frac{dy}{dx} = y\,\frac{10x^2 + 22x + 6}{(2x+1)(3x-1)(4x+3)}$$

$$= \frac{(2x+1)^{\frac{1}{2}}(3x-1)^{\frac{2}{3}}}{(4x+3)^{\frac{3}{4}}}\,\frac{10x^2 + 22x + 6}{(2x+1)(3x-1)(4x+3)}$$

$$\frac{dy}{dx} = \frac{2(5x^2 + 11x + 3)}{(2x+1)^{\frac{1}{2}}(3x-1)^{\frac{1}{3}}(4x+3)^{\frac{7}{4}}}$$

EXERCISE 6

1 Find the derivatives of the following functions.

a $f(x) = 5^{2x}$ **b** $f(x) = (x + 1)^{x-1}$ **c** $f(x) = e^{\sin^2 x}$

d $f(x) = 3^{e^x}$ **e** $f(x) = (\cos x)^x$

2 Find the derived function for each of these functions.

a $f(x) = x^{x^2}$ **b** $f(x) = \pi^{x^3}$ **c** $f(x) = \dfrac{e^x \sin x}{x}$

d $f(x) = xe^{-x}\cos x$ **e** $f(x) = (1 - x^3)^{\sin x}$

3 Show that, for $y = e^{\cos 2x \cos^2 x}$, $\dfrac{dy}{dx} = -2\cos x \sin 3x\, e^{\cos 2x \cos^2 x}$

4 Show that $y = e^{\cos^2 x} \Rightarrow \dfrac{d^2y}{dx^2} = (\sin^2 2x - 2\cos 2x)y$.

5 Show that the tangent to the curve with equation $y = (\sin x)^x$ at $x = \dfrac{\pi}{2}$ is parallel to the x-axis.

6 For the curve with equation $y = x^{\sin x}$ show that the tangent at $x = \dfrac{\pi}{2}$ is inclined at $\dfrac{\pi}{4}$ to the x-axis.

7 Find the equation of the tangent to the curve with equation $y = \dfrac{x(2x+1)^{\frac{3}{2}}}{(3x-4)^{\frac{2}{3}}}$ at the point where $x = 4$.

8 Find the gradient of the curve with equation $y = \dfrac{(x+3)^{\frac{1}{2}}}{x(x+2)^{\frac{1}{3}}}$ at the point where $x = 6$.

9 Find the value of x for which the graph of $y = \dfrac{x^{\frac{1}{2}}(3-x)^{\frac{1}{6}}}{(2x+1)^{\frac{2}{3}}}$ is stationary.

10 Differentiate: **a** $\dfrac{(x+1)^{\frac{1}{2}}(x-1)^{\frac{1}{3}}}{(x+2)^{\frac{1}{4}}}$ **b** $\dfrac{(2x+3)^{\frac{5}{2}}}{x(x-1)^{\frac{2}{3}}}$

Parametric equations

Although we may have a function $y = f(x)$, it is sometimes more convenient to express the variables x and y as functions of a third variable, say t. Equations $x = x(t)$ and $y = y(t)$ are called *parametric equations* and t is referred to as the *parameter.*

Each value of t will uniquely define a point $(x(t), y(t))$ on the curve $y = f(x)$.

For example, the equation of the line illustrated here is $3x + 4y = 24$.

The line can also be represented by the parametric equations

$$x = 4(1 + t)$$
$$y = 3(1 - t)$$

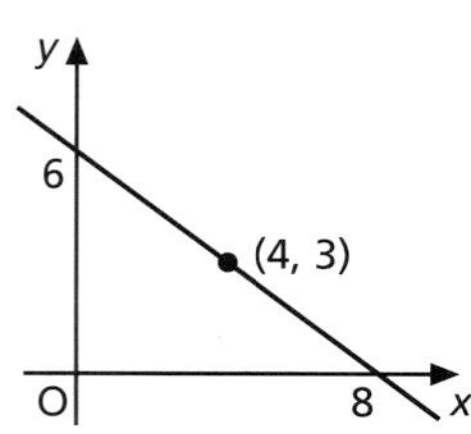

Note that $t = 1$ gives rise to the point $(8, 0)$ and $t = -1$ corresponds to the point $(0, 6)$.

It is also possible to eliminate a parameter t from a pair of equations.
Given $x = x(t)$ and $y = y(t)$, we get, from the first equation, $t = x^{-1}(x)$.
Substituting in the second equation gives $y = y(x^{-1}(x))$.
This equation is referred to as the *constraint equation* of the function.

We can demonstrate this on the pair of parametric equations above.

$$x = 4 + 4t \text{ and } y = 3 - 3t$$

$$\Rightarrow \quad t = \frac{(x-4)}{4}$$

$$\Rightarrow \quad y = 3 - 3\frac{(x-4)}{4}$$

$$\Rightarrow \quad 4y = 12 - 3x + 12$$

$$\Rightarrow \quad 4y + 3x = 24$$

Note that the parametric representation of a curve is not unique. For any function $y = f(x)$, we can define a parameter u such that $x = k(u)$, and thus we can find a pair of equations $x = k(u)$ and $y = f(k(u))$.

For example, the pair

$$x = 4(3 + u)$$
$$y = -3(1 + u)$$

also represents the above line.

Parametric equations of the circle

Consider the pair of equations

$$x = 4 + r\cos\theta$$
$$y = 3 + r\sin\theta$$

where r is a constant. To eliminate the parameter θ

$$r\cos\theta = x - 4 \Rightarrow r^2\cos^2\theta = (x-4)^2$$
$$r\sin\theta = y - 3 \Rightarrow r^2\sin^2\theta = (y-3)^2$$
$$\Rightarrow (x-4)^2 + (y-3)^2 = r^2\cos^2\theta + r^2\sin^2\theta = r^2[\cos^2\theta + \sin^2\theta]$$
$$\Rightarrow (x-4)^2 + (y-3)^2 = r^2$$

which represents the circle centre (4, 3) radius r.

From the diagram it is clear that the parameter θ can be interpreted as the amount the radius rotates anticlockwise from the x-direction.
The constraint equation of the circle is

$$(x-4)^2 + (y-3)^2 = r^2$$

and the parametric equations (also known as *freedom equations*) are

$$x = 4 + r\cos\theta$$
$$y = 3 - r\sin\theta$$

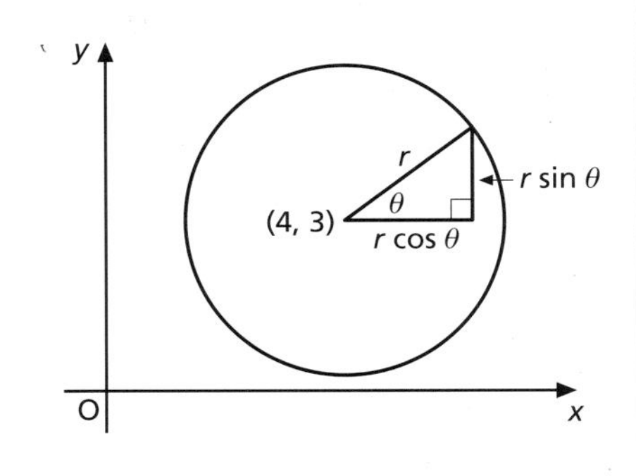

Example 1 **a** Find the constraint equation of the locus defined by the parametric equations $x = t + \frac{1}{t}$ and $y = t - \frac{1}{t}$.
b Sketch the curve.

a t can be eliminated by squaring both equations and subtracting them.

$$x^2 = t^2 + 2 + t^{-2}$$
$$y^2 = t^2 - 2 + t^{-2}$$
$$\Rightarrow x^2 - y^2 = 4$$

b
- $y = 0 \Rightarrow x = \pm 2$; $x = 0 \Rightarrow y^2 = -4 \Rightarrow y \notin R$ so the curve does not intersect the y-axis.
- Using implicit differentiation on the constraint equation

$$2x - 2y\frac{\mathrm{d}y}{\mathrm{d}x} = 0$$
$$\Rightarrow \quad \frac{\mathrm{d}y}{\mathrm{d}x} = \frac{x}{y}$$
$$\frac{\mathrm{d}y}{\mathrm{d}x} = 0 \Rightarrow x = 0 \Rightarrow y \notin R$$ So the curve has no stationary points.

- If $y = 0$, then $\frac{\mathrm{d}y}{\mathrm{d}x}$ is undefined. So, at $x = \pm 2$, the tangents to the curve are parallel to the y-axis.
- $y = \pm\sqrt{(x^2 - 4)} \Rightarrow x^2 - 4 \geq 0 \Rightarrow x \leq -2$ or $x \geq 2$
- $y = \pm\sqrt{(x^2 - 4)}$ means that, as $x \to \infty$, $y \to \pm\sqrt{(x^2)}$. So $y = \pm x$ are non-vertical asymptotes.

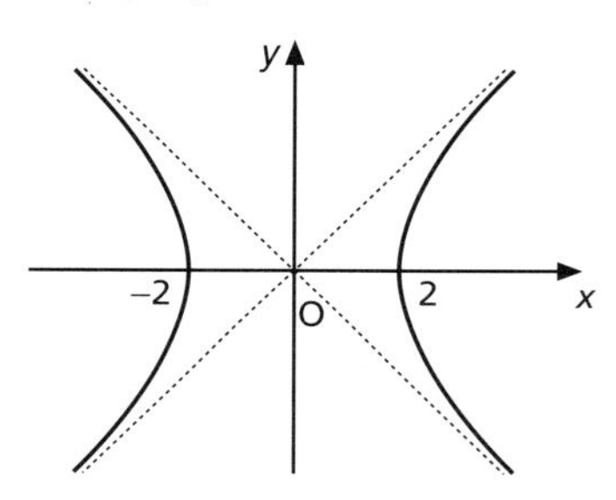

The graph can be verified:
- using a graphics calculator (most can now be set to draw parametric representatives directly)
- using a spreadsheet.

	A	B	C
1	parameter	x-value	y-value
2	-10	=A2+1/A2	=A2-1/A2
3	=A2+0.2	=A3+1/A3	=A3-1/A3

EXERCISE 7A

1 For each of the following pairs of parametric equations find the corresponding constraint equation.

a $x = 1 - t$, $y = 1 + 2t$

b $x = 3t$, $y = \frac{3}{t}$

c $x = 5p^2$, $y = 10p$

d $x = 4 \cos \theta$, $y = 3 \sin \theta$

e $x = 5 \sec \theta$, $y = 12 \tan \theta$

f $x = 3 + 2 \sin \theta$, $y = 2 - 3 \cos \theta$

g $x = \cos \theta + \sin \theta$, $y = \cos \theta - \sin \theta$

h $x = \frac{2 + 3t}{4}$, $y = \frac{3 - 4t}{5}$

Hint
When trigonometric ratios are involved, try squaring both sides of each equation.

2 Sketch the locus represented by each of the above pairs of parametric equations.

A diversion

EXERCISE 7B

1 Use a spreadsheet or graphics calculator to help you sketch the following loci.

a $x = t^2;\ y = \dfrac{1}{t}$

b $x = e^t;\ y = e^{2t}$

c $x = t^2 + t;\ y = t^3$

d $x = 2\cos\theta + 3\sin\theta;\ y = 3\cos\theta - 2\sin\theta$

2 Some very aesthetic curves can be drawn using parametric equations. Explore the following using a spreadsheet or graphics calculator.

a The cardioid: $x = 2\cos\theta + 2\cos^2\theta;\ y = 2\sin\theta + \sin 2\theta$

b The loopless limaçon: $x = 4\cos\theta + 2\cos^2\theta;\ y = 4\sin\theta + \sin 2\theta$

c The limaçon: $x = 2\cos\theta + 4\cos^2\theta;\ y = 2\sin\theta + 2\sin 2\theta$

d The Archimedean spiral: $x = \theta\cos n\theta,\ y = \theta\sin n\theta$, where n determines the number of turns

e The lobiates: $x = \cos\theta\cos n\theta;\ y = \sin\theta\cos n\theta$, where n determines the number of lobes.

First and second derivatives of parametric equations

When $x = x(t)$, $y = y(t)$ and $y = f(x)$ then $f'(x)$ and $f''(x)$ can be determined by making use of the chain rule.

$$f'(x) = \frac{dy}{dx} = \frac{dy}{dt} \times \frac{dt}{dx}$$

$$= \frac{dy}{dt} \div \frac{dx}{dt}$$

$$\boxed{f'(x) = \frac{y'(t)}{x'(t)}}$$

$$f''(x) = \frac{d^2y}{dx^2} = \frac{d}{dx}\left(\frac{dy}{dx}\right) = \frac{d}{dt}\left(\frac{dy}{dx}\right) \times \frac{dt}{dx}$$

$$= \frac{d}{dt}\left(\frac{y'(t)}{x'(t)}\right) \times \frac{1}{x'(t)}$$

$$= \frac{x'(t)y''(t) - x''(t)y'(t)}{(x'(t))^2} \times \frac{1}{x'(t)}$$

$$\boxed{f''(x) = \frac{x'(t)y''(t) - x''(t)y'(t)}{(x'(t))^3}}$$

Example 1 Find $f'(x)$ and $f''(x)$ when $x = 4 + 4t$ and $y = 3 - 3t^2$.

$x'(t) = 4;\quad y'(t) = -6t$

$$\Rightarrow\ f'(x) = \frac{-6t}{4}$$

$x''(t) = 0;\quad y''(t) = -6$

$$\Rightarrow\ f''(x) = \frac{4 \times (-6) - 0 \times (-6t)}{(4)^3} = -\frac{3}{8}$$

Example 2 Given the locus represented by the parametric equations

$$x = 3 + 5\cos\theta$$
$$y = 4 + 5\sin\theta$$

a find the point on the locus corresponding to a parameter value $\theta = \frac{\pi}{2}$,
b find the gradient of the tangent,
c find the concavity at this point.

a $$x\left(\frac{\pi}{2}\right) = 3 + 5\cos\frac{\pi}{2} = 3$$
$$y\left(\frac{\pi}{2}\right) = 4 + 5\sin\frac{\pi}{2} = 9$$

The point on the locus is (3, 9).

b $x'(\theta) = -5\sin\theta$; $y'(\theta) = 5\cos\theta$

$$\Rightarrow f'(x) = \frac{5\cos\theta}{-5\sin\theta} = -\cot\theta$$

When $\theta = \frac{\pi}{2}$, $f'(x) = -\cot\frac{\pi}{2} = 0$.

The tangent is horizontal.

c $x''(\theta) = -5\cos\theta$; $y''(\theta) = -5\sin\theta$

$$\Rightarrow f''(x) = \frac{-5\sin\theta\,(-5\sin\theta) - (-5\cos\theta)\,5\cos\theta}{(-5\sin\theta)^3} = \frac{25(\sin^2\theta + \cos^2\theta)}{(-5\sin\theta)^3}$$
$$= \frac{-1}{5\sin^3\theta}$$

When $\theta = \frac{\pi}{2}$, $f''(x) = -\frac{1}{5}$.

$f''(x) < 0$ so the curve is concave down.

EXERCISE 8A

1 Find $\frac{dy}{dx}$ and $\frac{d^2y}{dx^2}$ for the curve defined by each of the following pairs of parametric equations.

a $x = t$, $y = \frac{1}{t}$

b $x = t^2$, $y = \ln t$

c $x = t + \sin t$, $y = t - \cos t$

d $x = 3t^3 - t$, $y = 4t^2$

e $x = \theta - \sin\theta$, $y = 1 + \cos\theta$

2 A curve is defined by the equations $x = t^2 + \frac{2}{t}$ and $y = t^2 - \frac{2}{t}$.

a Find the coordinates of the turning point on the curve.
b Establish the nature of the turning point by considering the concavity of the curve.

3 a Find $\frac{dy}{dx}$ and $\frac{d^2y}{dx^2}$ for the curve defined by $x = t^2 - \frac{1}{t^2}$, $y = t^2 + \frac{1}{t^2}$.
b Determine the coordinates of the turning point on the curve.
c Prove that the curve is always concave up.

4 For the curve defined by $x = \frac{2t}{1 - t^2}$, $y = \frac{1 + t^2}{1 - t^2}$, show that

a $\frac{dy}{dx} = \frac{x}{y}$ b $\frac{d^2y}{dx^2} = \frac{1}{y^3}$

5 A spiral is generated by the parametric equations $x = e^{\theta} \sin \theta$, $y = e^{\theta} \cos \theta$. Find the coordinates and the nature of the turning points which occur when $0 \le \theta \le 2\pi$.

6 A curve is defined by $x = 2 \sin \theta + \cos 2\theta$, $y = 2 \cos \theta - \sin 2\theta$, $0 \le \theta \le 2\pi$.
a Show that, at the points where $\frac{dy}{dx} = 0$, $\left|\frac{d^2y}{dx^2}\right| = \frac{\sqrt{3}}{4}$.
b Find the points where the gradient is undefined.

EXERCISE 8B

1 A curve is defined by $x = (1 + t)^{\frac{1}{2}}$, $y = (1 - t)^{-\frac{1}{2}}$.
a Show that this graph has only one critical point and determine its nature.
b The curve has a vertical asymptote. Identify it by finding where the gradient is undefined.

2 $x = 1 + \sin^2 \theta$ and $y = 1 - \sec^2 \theta$ are the parametric representations of a curve. Show that, at the point where $\tan \theta = 2$, the equation of the tangent is $25x + y = 41$.

3 An *astroid* is defined by $x = \sin^3 t - \cos^3 t$, $y = \sin^3 t + \cos^3 t$.
a Show that the equation of the tangent to the astroid at the point where $t = \tan^{-1} \frac{3}{4}$ is $5x + 35y = 24$.
b The curve cuts the axes where the gradient is zero or undefined. Calculate the points of intersection of the curve with the axes.

4 a Find, in terms of θ, $\frac{dy}{dx}$ and $\frac{d^2y}{dx^2}$ for the curve defined by: $x = 1 + \sin \theta$
$y = \theta + \sin \theta \cos \theta$
b Show that the constraint equation of this curve is
$$y = \sin^{-1}(x - 1) + (x - 1)\sqrt{(2x - x^2)}.$$
c Find $\frac{dy}{dx}$ and $\frac{d^2y}{dx^2}$ by direct differentiation and verify that your results agree with your answers in part **a**.

5 a Identify the turning points on the curve defined by $x = \frac{3t}{1 + t^3}$, $y = \frac{3t^2}{1 + t^3}$.
b Show that there are no points of inflexion.

CHAPTER 2.1 REVIEW

1 Given that $f(x) = \frac{1}{2}(x^3 - 5)$, find

a $f'(x)$ and $f^{-1}(x)$

b the derivative of $f^{-1}(x)$, using the rule $\dfrac{\mathrm{d}f^{-1}(x)}{\mathrm{d}x} = \dfrac{1}{f'(f^{-1}(x))}$

2 Given that $y = (1 - x^2)^{\frac{1}{2}} + 4 \sin^{-1} x$, find $\dfrac{\mathrm{d}x}{\mathrm{d}y}$.

3 Differentiate $\sin^{-1} 5x$.

4 Find the gradient of the tangent to the curve $y = \ln \tan^{-1}(3x)$ where $x = \dfrac{1}{\sqrt{3}}$.

5 Differentiate $e^{\frac{x}{2}} \cos^{-1} 2x$.

6 Differentiate $\tan^{-1}\left(\dfrac{1 + \cos x}{\sin x}\right)$.

7 Given $\dfrac{1}{x} + \dfrac{1}{y} = \dfrac{1}{\pi}$, find $\dfrac{\mathrm{d}y}{\mathrm{d}x}$ and $\dfrac{\mathrm{d}^2y}{\mathrm{d}x^2}$.

8 Given $y = \dfrac{2^x}{x + 2}$, find the rate of change of y with respect to x, when $x = 2$.

9 Find the constraint equation of the curve defined parametrically by

$$x = 2 + \frac{1}{t}, \quad y = \frac{t^2 + 1}{t^2 - 1}$$

10 A curve is defined parametrically by $x = 2t - 3 - \dfrac{1}{t}$, $y = t - 1 - \dfrac{2}{t}$.

a **(i)** Find the equations of the tangents to the curve at the points where it crosses the x-axis.

(ii) Find the coordinates of the point of their intersection.

b Show that there are no points of inflexion on this curve.

CHAPTER 2.1 SUMMARY

1 The *derivative of the inverse* of a function can be found using the rule

$$\frac{df^{-1}(x)}{dx} = \frac{1}{f'(f^{-1}(x))}$$

2 The *reciprocal derivative*: a useful result is $\frac{dx}{dy} = \frac{1}{\frac{dy}{dx}}$.

3 The *derivatives of the inverse trigonometric functions* are:

$$\sin^{-1} x = \frac{1}{\sqrt{(1 - x^2)}}, \qquad \cos^{-1} x = \frac{-1}{\sqrt{(1 - x^2)}}, \qquad \tan^{-1} x = \frac{1}{1 + x^2}$$

4 *Implicit and explicit functions*

(i) When the dependent variable, y, is expressed in terms of the independent variable, x (i.e. it is the subject of the formula), the function is said to be *explicit*. When the dependent variable is not the subject of the formula, then the function is said to be *implicit*.

(ii) The chain rule is used to differentiate terms which are functions of y.

$$\frac{d}{dx} g(y) = g'(y) \times \frac{dy}{dx}$$

5 *Logarithmic differentiation*

When a function is complicated by the occurrence of powers, roots, products and quotients of several factors it is useful to take logarithms before differentiating. For example

$$y = \frac{g(x)h(x)}{k(x)} \text{ becomes } \ln y = \ln g(x) + \ln h(x) - \ln k(x)$$

and thus

$$\frac{1}{y}\frac{dy}{dx} = \frac{g'(x)}{g(x)} + \frac{h'(x)}{h(x)} - \frac{k'(x)}{k(x)}$$

6 *Parametric equations*

(i) Given the function $y = f(x)$, it is sometimes more convenient to express the variables x and y as functions of a third variable, say t. Equations $x = x(t)$ and $y = y(t)$ are called *parametric equations* and t is referred to as the *parameter*.

(ii) Each value of t will uniquely define a point $(x(t), y(t))$ on the curve $y = f(x)$.

(iii) The equation $y = f(x)$ is often referred to as the *constraint equation* of the function.

7 *Differentiating using parameters*

$$f'(x) = \frac{y'(t)}{x'(t)}, \qquad f''(x) = \frac{x'(t)y''(t) - x''(t)y'(t)}{(x'(t))^3}$$

These results can be obtained by application of the chain rule.

2.2 Applications of Differentiation

Motion in a plane

Reminders

Motion in a line

Displacement, s, is often expressed as a function of time: $s(t)$.

Velocity, v, is the rate of change of displacement with respect to time: $v(t) = s'(t)$.

Acceleration, a, is the rate of change of velocity with respect to time:
$a(t) = v'(t) = s''(t)$.

Newton

It was mainly for its applications that Newton developed the calculus.

With reference to a suitable set of axes, a particle in motion on a plane is at position $(x(t), y(t))$ at a time t. Let the *displacement* of the particle from the origin be denoted by $s(t)$.

Displacement

The displacement, $s(t)$, is commonly represented by its position vector
$\mathbf{s}(t) = x(t)\boldsymbol{i} + y(t)\boldsymbol{j}$ where $\boldsymbol{i}$ and $\boldsymbol{j}$ are unit vectors in the x- and y-directions respectively.

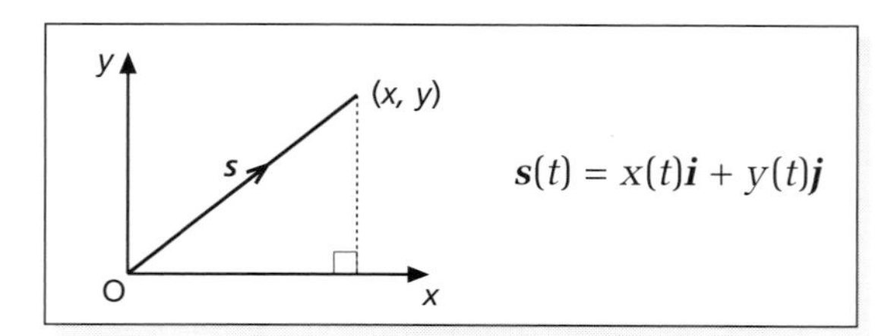

The distance from the origin is the magnitude of the displacement

$$|s| = \sqrt{((x(t))^2 + (y(t))^2)}$$

Velocity

The *velocity*, $\boldsymbol{v}$, of the particle is the rate of change of its displacement.

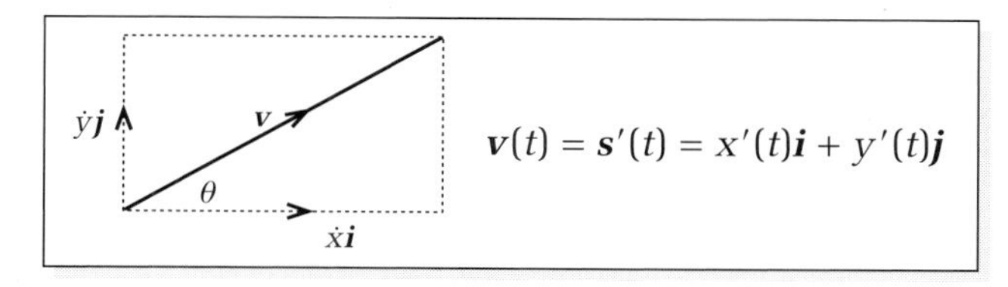

This is often shortened to

$$v = \dot{x}\boldsymbol{i} + \dot{y}\boldsymbol{j}$$

where the dot above the variable denotes differentiation with respect to t (time).

The *speed* of the particle is the magnitude of the velocity, namely $\sqrt{(\dot{x}^2 + \dot{y}^2)}$.

The *direction of motion* at any instant of time can be described with reference to the x-direction from the components of velocity:

$$\tan\theta = \frac{\dot{y}}{\dot{x}}$$

Note that in Leibniz notation this is

$$\frac{dy}{dt}\frac{dt}{dx} = \frac{dy}{dx}$$

Acceleration

The *acceleration*, $\boldsymbol{a}$, of the particle is the rate of change of its velocity with respect to time.

$$\boldsymbol{a}(t) = \boldsymbol{v}'(t) = \boldsymbol{s}''(t) = x''(t)\boldsymbol{i} + y''(t)\boldsymbol{j}$$

This is often written as $\boldsymbol{v} = \ddot{x}\boldsymbol{i} + \ddot{y}\boldsymbol{j}$.

The magnitude of the acceleration (which is not given a separate name) is $\sqrt{(\ddot{x}^2 + \ddot{y}^2)}$.

Example 1 The motion of a particle is modelled by the equations

$$x = 5t, \quad y = 5\sqrt{3}t - 5t^2$$

Find the speed of the particle and its direction of motion after 1 second.

Differentiating with respect to t

$$x'(t) = 5, \quad y'(t) = 5\sqrt{3} - 10t$$

$$t = 1 \Rightarrow x'(1) = 5, \quad y'(1) = 5\sqrt{3} - 10$$

$$\text{Speed} = \sqrt{(5^2 + (5\sqrt{3} - 10)^2)} = 5.2\,\text{m s}^{-1}$$

$$\tan\theta = \frac{5\sqrt{3} - 10}{5} \Rightarrow \theta = -0.262 \text{ radians}$$

$$\theta = -15^\circ \text{ or } (180 + (-15))^\circ = 165^\circ$$

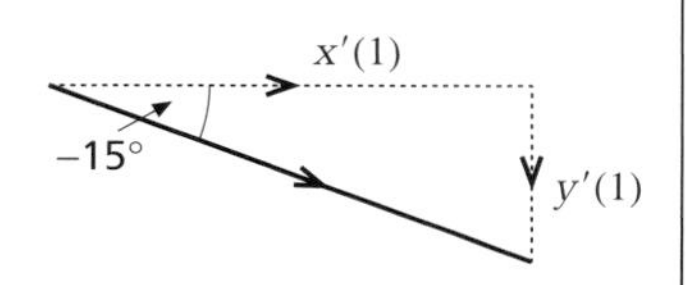

Noting that $x'(1) > 0$ and $y'(1) < 0$ tells us that it is the former.

Example 2 The equations of motion of a particle moving in a plane are

$$x = t^3 - 3t, \quad y = t^2 + t$$

where x, y are measured in metres and t in seconds. Find the magnitude and direction of

a the velocity,

b the acceleration after 3 s.

a Let the velocity have a magnitude $v\,\text{m s}^{-1}$ (or m/s) and a direction of $\alpha°$ to the x-direction.

$$x'(t) = 3t^2 - 3, \quad y'(t) = 2t + 1$$

$$\Rightarrow \quad x'(3) = 24, \quad y'(3) = 7$$

$$\Rightarrow \quad v = \sqrt{(24^2 + 7^2)} = 25\,\text{m s}^{-1}$$

Also $\tan \alpha° = \frac{7}{24} \Rightarrow \alpha° = 16.3°$ or $196.3°$

$x'(3) > 0$ and $y'(3) > 0 \Rightarrow \alpha° = 16.3°$

b Let the acceleration have a magnitude $a\,\text{m s}^{-2}$ and a direction of $\beta°$ to the x-direction.

$$x''(t) = 6t, \quad y''(t) = 2$$

$$\Rightarrow \quad x''(3) = 18, \quad y''(3) = 2$$

$$\Rightarrow \quad a = \sqrt{(18^2 + 2^2)} \approx 18.1\,\text{m s}^{-2}$$

$\tan \beta° = \frac{2}{18} \Rightarrow \beta° = 6.3°$ or $186.3°$

$x''(3) > 0, \quad y''(3) > 0 \Rightarrow \beta° = 6.3°$

EXERCISE 1

Note

Unless otherwise stated, t represents time measured in seconds and x and y represent distances measured in metres.

Remember also that for problems involving trigonometric ratios the calculator should be set to work with radian measure.

1 A particle moving in a plane is governed by the laws

$$x = t^2 + 3\sin t, \quad y = 2t^2 - \cos t$$

Find, when $t = 10\pi$,

a the speed of the particle,

b the direction of motion relative to the x-direction.

2 The equations of motion of a particle moving in a plane are

$$x = 2t^3 + 3t^2, \quad y = 3t^2 + 6t$$

Find, when **(i)** $t = 2$, **(ii)** $t = 3$,

a the position of the particle,

b the magnitude and direction of its velocity,

c the magnitude and direction of its acceleration.

3 The motion of a particle moving in a plane can be described by

$$x = 3 + \ln(t + 2), \quad y = 4 + \ln(t + 3)$$

Find, when **(i)** $t = 0$, **(ii)** $t = 1$,

a the position of the particle,

b the magnitude and direction of its velocity,

c the magnitude and direction of its acceleration.

4 As it takes off, a light aircraft's movement can be modelled by the equations

$$x = t \ln t, \quad y = t^2 \ln t$$

Find, when $t = 10$,

a the magnitude and direction of its velocity,

b the magnitude and direction of its acceleration.

5 **a** Find when a particle is instantaneously at rest (i.e. the instant when its speed is zero) if it moves in a plane subject to the equations of motion

$$x = 3t^2 - 6t, \quad y = t^3 - 3t$$

b Find also, at that instant,

(i) its position,

(ii) the magnitude and direction of its acceleration.

6 **a** Find when a particle is instantaneously at rest if it moves in a plane governed by the laws

$$x = 2t^3 - 6t^2, \quad y = 5t^2 - 20t$$

b Find also where it is and the magnitude and direction of its acceleration at that instant.

7 A particle moves in a plane in accordance with the equations of motion

$$x = e^t + e^{-t}, \quad y = e^t - e^{-t}$$

a Express, in terms of e^{2t}, **(i)** the speed, **(ii)** the tangent of the angle that the direction of motion makes with the x-direction.

b Discuss the behaviour of the speed and the angle as t increases.

8 **a** Show that, for a particle moving in a plane such that

$$x = \sin^{-1}\left(\frac{t}{10}\right) \text{ and } y = \cos^{-1}\left(\frac{t}{10}\right)$$

the velocity and acceleration share a common fixed direction.

b Find the speed and acceleration when $t = 6$.

9 A particle is moving in a plane under the equations of motion

$$x = 1 + \cos t - \sin t, \quad y = 1 - \cos t + \sin t$$

If v and f are the velocity and acceleration at any instant, prove that $v^2 + f^2 = 4$.

Related rates of change

Reminders

- When one variable, x, is a function of another, u, then $\dfrac{dx}{du} = \dfrac{1}{\dfrac{du}{dx}}$.
- When y is a function of x, say $y = f(x)$, and both y and x are functions of a third variable, u, such that $x = x(u)$ and $y = y(u)$, then the rates of change of x and y with respect to u are related to the rate of change of y with respect to x by the chain rule.

  $$\frac{dy}{dx} = \frac{dy}{du} \times \frac{du}{dx} = \frac{dy}{du} \div \frac{dx}{du}$$

  This fact can be used in solving many problems.

Example 1 A spherical balloon is being inflated at a constant rate of 240 cm^3 per second.

a At what rate is the radius increasing when it is equal to 8 cm?
b At what rate is the radius increasing after 5 seconds?

Let $V(t)$ represent the volume after t seconds. Let $r(t)$ represent the corresponding radius.

The question gives us that $\dfrac{dV}{dt} = 240$. The question asks us for $\dfrac{dr}{dt}$.

By the chain rule we know $\dfrac{dr}{dt} = \dfrac{dr}{dV} \times \dfrac{dV}{dt}$.

We must find $\dfrac{dr}{dV}$, so find a relation between r and V.

$$V = \frac{4}{3}\pi r^3 \Rightarrow \frac{dV}{dr} = 4\pi r^2 \Rightarrow \frac{dr}{dV} = \frac{1}{4\pi r^2}$$

Thus

$$\frac{dr}{dt} = \frac{1}{4\pi r^2} \times 240 = \frac{60}{\pi r^2}$$

a When $r = 8$, the radius is growing at $\dfrac{60}{64\pi} = 0.3$ cm per second (to one decimal place).

b After 5 seconds the volume of the sphere is $240 \times 5 = 1200$ cm^3.
The corresponding radius, using the formula $V = \frac{4}{3}\pi r^3$, is 6.59 cm. Thus

$$\frac{dr}{dt} = \frac{60}{\pi(6.59)^2} = 0.4 \text{ cm per second (to one decimal place)}$$

Reminder

When the relation between two variables is stated implicitly and the two are related to a third, the relation between the associated rates can again be used to solve problems.

Example 2 A particle is moving in a circle of radius 5 m and centre the origin. When the particle is at the point $(2.5\sqrt{3}, -2.5)$, the rate of change of the x-coordinate with respect to time is 25 m/s. Find the rate of change of the y-coordinate.

The equation of the locus of the particle is $x^2 + y^2 = 25$. Differentiating this equation implicitly with respect to t (time) gives

$$2x\frac{dx}{dt} + 2y\frac{dy}{dt} = 0$$

The question gives $\frac{dx}{dt} = 25$ so when $x = 2.5\sqrt{3}$ and $y = -2.5$

$$5\sqrt{3} \times 25 + (-5) \times \frac{dy}{dt} = 0 \Rightarrow \frac{dy}{dt} = 25\sqrt{3} \text{ m/s}$$

The y-coordinate is changing at the rate of $25\sqrt{3}$ metres per second.

EXERCISE 2A

1 A labourer is digging a track and removes $V = \left(t - \frac{t^2}{14\,400}\right)$ m^3 of earth in t hours.
The track is 5 metres wide and is being excavated to a depth of 10 cm.
The excavated region can be considered as a cuboid.

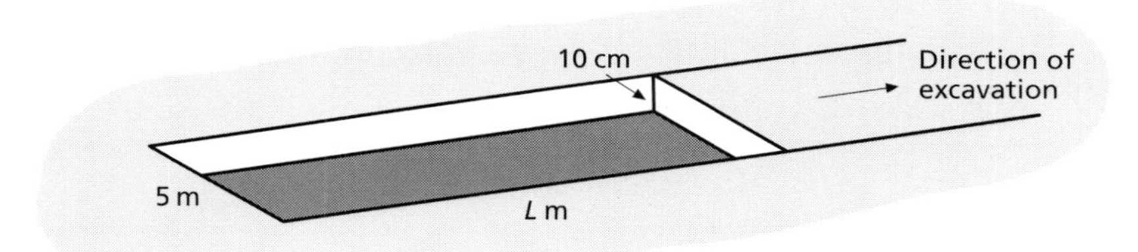

a Write down an expression for the volume of earth removed in terms of L, the length of the trench in metres.
b Find an expression for the rate of change of volume with respect to length.
c Work out an expression for $\frac{dV}{dt}$.
d Find an expression for the rate of change of the length of the track with respect to time.
e At what rate is the labourer removing the earth after half an hour:
(i) in terms of m^3/h **(ii)** in terms of m/h?

2 A particle moves along the curve $y = x^3$ so that the x-component of its velocity v is always 4 m/s.
a Differentiate with respect to time, t, to obtain a relationship between $\frac{dy}{dt}$ and $\frac{dy}{dx}$.
b Calculate the y-component of its velocity when **(i)** $x = 3$ **(ii)** $y = -1$.

3 Given that $P = (4x - 3)^7$ and $Q = 7x(2x - 3)$, express $\frac{dP}{dQ}$ in terms of x.

4 Oil is dripping from a car engine on to a garage floor making an ever-increasing circular oil stain. The radius, r cm, of this stain is increasing at a constant rate of 1 cm per hour.

  a State the value of $\frac{dr}{dt}$, the rate of change of radius with time.

  b **(i)** State the expression which relates the radius to the area, $A\,\text{cm}^2$, of a circle.

  (ii) Hence, state $\frac{dA}{dr}$, the rate of change of area with respect to radius.

  c Find the rate at which the area of the stain is increasing with time, when the radius is 4 cm.

5 A metal cube of edge x cm, is heated. Each edge expands at a rate of 0.002 cm per minute.

  a **(i)** Express the volume, $V\,\text{cm}^3$, as a function of x.

  (ii) Hence find an expression for the rate of change of volume with respect to edge length.

  b Find the rate at which the volume of the cube is expanding with respect to time when the edge is 10 cm long.

6 A spherical balloon is being inflated. Its volume, $V\,\text{cm}^3$, is increasing at the rate of $\frac{24\pi}{5}\,\text{cm}^3$ per second.

  a Find an expression which gives the rate of change of the volume of the sphere, $V\,\text{cm}^3$, with respect to its radius.

  b Find the rate at which the radius is increasing with respect to time when the volume is $\frac{32\pi}{3}\,\text{cm}^3$.

Harder questions

EXERCISE 2B

1 A disc starts from rest and is speeded up. The number of revolutions per minute that the disc makes increases at a constant rate. After t minutes it has rotated through $k = (0.5t^2 + t)$ revolutions.

  a Find $\frac{dk}{dt}$, the speed of the wheel in revolutions per minute (r.p.m.).

  b How long will it take to achieve a speed of 16 r.p.m.?

  c A water droplet on the disc moves out from the centre at a constant rate of 2 mm/s. At how many millimetres per revolution is the droplet moving out after **(i)** 1 minute **(ii)** 30 seconds?

2 According to Boyle's law the volume, $V\,\text{m}^3$, of a fixed mass of gas is inversely proportional to the pressure, P newtons/m^2, of the gas.
In one particular case this leads to the relationship, $PV = 400$. The rate of change of the volume of the gas has been measured as $12\,\text{m}^3/\text{s}$. Find the rate of change of pressure with respect to time when the volume of the gas is $40\,\text{m}^3$.

3 A conical container of height 12 cm and base radius 3 cm is being filled with liquid at a rate of 25 cm^3 per minute.
At time t, the surface of the liquid is a circle of radius $r(t)$ and the depth of the liquid is $h(t)$.

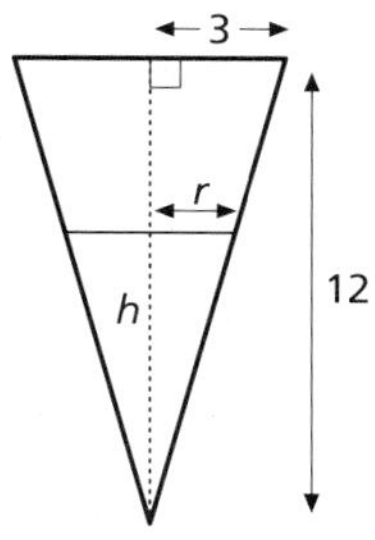

a **(i)** Using similar triangles, express h in terms of r.

(ii) Hence find an expression in terms of r alone for the volume, V cm^3, of liquid in the container after t minutes.

(iii) Find an expression for $\dfrac{dV}{dr}$.

b Find the rate at which the radius is increasing, in centimetres per minute, when the depth of liquid is 5 cm.

c Find the rate of increase of the depth when the liquid is 4 cm deep.

4 According to Charles' law, the volume V m^3 of a fixed mass of gas is directly proportional to its absolute temperature, T °A (degrees Absolute).
In one particular case this is expressed as $V = 20T$. If the temperature is rising at the rate of 10 °A per minute, find the rate at which the volume is increasing.

5 Grain is being ejected from a chute at the rate of 1.5 m^3 per minute and is forming a conical heap whose height is three quarters of its base diameter.
Find the rate of increase of the height of this cone when the base has a diameter of 3 metres.

6 A liquid is being poured in to a cone of base radius 30 cm and height 40 cm at a rate of 10π cm^3 per second.
Find, at the instant the circular surface has a radius of 10 cm, the rate of change of:

a the radius of the circular surface,

b the depth of the liquid,

c the area of the circular surface.

7 Show that, for a particle moving in a straight line with displacement s, velocity v and acceleration a,

$$a = v\frac{dv}{ds} = \frac{d}{ds}\left(\frac{v^2}{2}\right)$$

8 A particle describes the locus defined by the ellipse $4x^2 + 9y^2 = 36$.
If at the point $(1.5, \sqrt{3})$ the y-coordinate of the particle is increasing at the rate of 10 cm/s, calculate the rate of change, in cm/s, of the x-coordinate.

9 A particle moves along the curve with equation $x^2 - 3xy + 2y^2 = 4$.
If at the point (2, 3) the rate of change of the y-coordinate is 3 units per second, calculate the rate of change of the x-coordinate with respect to time.

10 A metal cylinder is being rolled in a steel press. It maintains its cylindrical shape and volume but gets longer and thinner.
When the radius of the cylinder is 3 cm and its length is 12 cm, the rate of decrease of the radius is 0.2 cm/s. Find the rate of increase of the length with respect to time at that instant.

11 A rocket takes off vertically. The angle of elevation, z radians, of the rocket when it is at a height of x metres above the launch pad is given by $z = \sin^{-1}\left(\frac{x}{y}\right)$, where y is the direct distance in metres between the rocket and the observer.

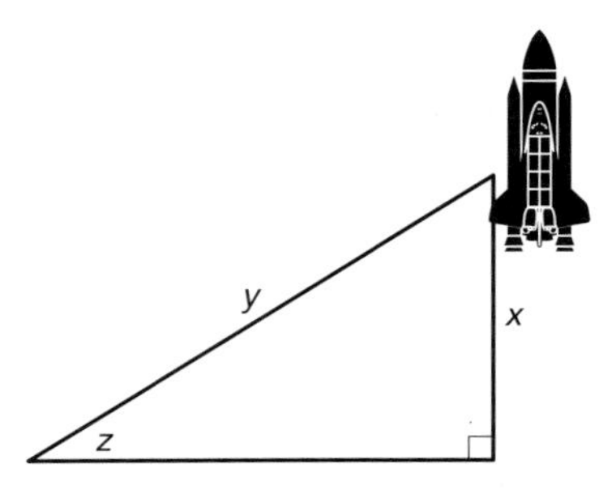

When $x = \frac{8}{5}$ and $y = \frac{5}{2}$, $\frac{dx}{dt} = 10$.

a Find, for these values, $\frac{dy}{dt}$, the speed at which the observer sees the rocket moving away.
b Find, at this instant, the rate at which the angle Z is changing.

CHAPTER 2.2 REVIEW

1 A particle moves in a plane satisfying the laws of motion

$$x = 4t^3, \quad y = 3t^4.$$

When $t = 2$, find

a the magnitude of the velocity,
b its direction with respect to the x-direction,
c the magnitude of the acceleration,
d its direction with respect to the x-direction.

2 A grain hopper is in the shape of an inverted square pyramid of height 10 units and side of base 5 units. Grain is flowing into it at the rate of 8 units^3 per minute. When the depth, h, of the grain is 2 units, calculate the rate of increase with respect to time of:

a the depth of grain,
b the area of the square top surface of the grain.

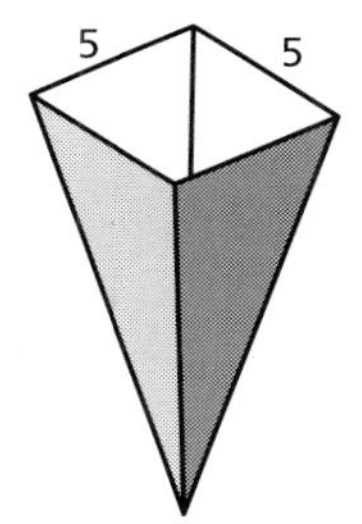

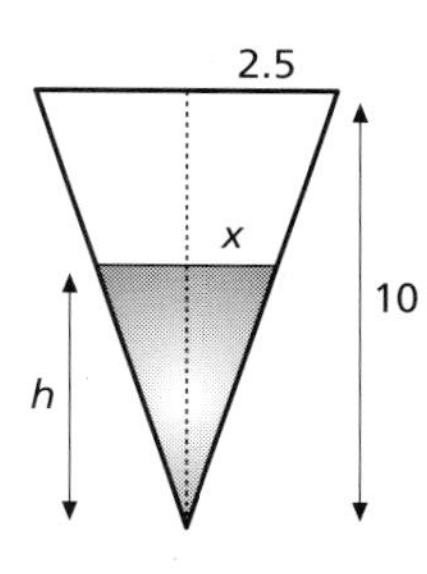

CHAPTER 2.2 SUMMARY

1 **(i)** A point (x, y) moves in a plane. Its motion in a plane is usually resolved into two components: displacement in the x-direction and displacement in the y-direction. These components are expressed as functions of time, t.

(ii) The displacement, s, has components $x = x(t)$ and $y = y(t)$.

(iii) The *distance from the origin*, $|s|$, is the magnitude of the displacement: $|s| = \sqrt{((x(t))^2 + (y(t))^2)}$.

(iv) The *direction measured from x-axis*, θ is given by: $\tan\theta = \dfrac{y(t)}{x(t)}$.

(Ambiguities are resolved by considering the components of displacement.)

(v) The *velocity*, v, has components $x'(t)$ and $y'(t)$.

(vi) The *speed*, $|v|$, is the magnitude of the velocity: $|v| = \sqrt{((x'(t))^2 + (y'(t))^2)}$.

(vii) The *direction of motion*, α, is given by $\tan\alpha = \dfrac{y'(t)}{x'(t)}$.

(Ambiguities are resolved by considering the components of velocity.)

(viii) The *acceleration*, a, has components $x''(t)$ and $y''(t)$.

(ix) The *magnitude of acceleration*: $|a| = \sqrt{((x''(t))^2 + (y''(t))^2)}$

(x) The *direction of acceleration*, β, is given by $\tan\beta = \dfrac{y''(t)}{x''(t)}$.

(Ambiguities are resolved by considering the components of acceleration.)

2 Rates of change are related through:

(i) the chain rule

$$\frac{dy}{dx} = \frac{dy}{dt} \times \frac{dt}{dx}$$

(ii) the reciprocal rule

$$\frac{dy}{dx} = \frac{1}{\frac{dx}{dy}}$$

(iii) implicit differentiation with respect to a third parameter.

3 Further Integration

Historical note

Jacques Bernoulli (1654–1705)

The development of calculus over the centuries has been the work of many people – with Euler, Newton and Leibniz predominant. One family in particular, the Bernoulli family, made many contributions in calculus.

Jacques Bernoulli was one of the first to make the study of calculus popular on the Continent while Johan Bernoulli, his younger brother, developed an integral calculus and introduced the term 'integral'.

Johan Bernoulli (1667–1748)

Correspondence between Euler, Johan and Daniel, his son, were instrumental in the development of solutions to differential equations.

Inverse trigonometric functions and standard integrals

Integrals of $\frac{1}{\sqrt{1-x^2}}$ and $\frac{1}{1+x^2}$

$$\frac{d}{dx}(\sin^{-1} x) = \frac{1}{\sqrt{1-x^2}}, \text{ so } \sin^{-1} x \text{ is an antiderivative of } \frac{1}{\sqrt{1-x^2}}$$

$$\frac{d}{dx}(\tan^{-1} x) = \frac{1}{1+x^2}, \text{ so } \tan^{-1} x \text{ is an antiderivative of } \frac{1}{1+x^2}$$

Hence we have

$$\int \frac{dx}{\sqrt{1-x^2}} = \sin^{-1} x + c \quad \text{and} \quad \int \frac{dx}{1+x^2} = \tan^{-1} x + c$$

Integrals of $\dfrac{1}{\sqrt{a^2 - x^2}}$ *and* $\dfrac{1}{a^2 + x^2}$

Using the substitution $x = at$ we get $\mathrm{d}x = a\,\mathrm{d}t$

$$\begin{aligned}\int \frac{\mathrm{d}x}{\sqrt{a^2 - x^2}} &= \int \frac{a}{\sqrt{a^2 - a^2t^2}}\,\mathrm{d}t \\ &= \int \frac{a}{\sqrt{a^2(1 - t^2)}}\,\mathrm{d}t \\ &= \int \frac{a}{a\sqrt{1 - t^2}}\,\mathrm{d}t \\ &= \int \frac{\mathrm{d}t}{\sqrt{1 - t^2}} \\ &= \sin^{-1} t + c \\ &= \sin^{-1}\left(\frac{x}{a}\right) + c\end{aligned}$$

$$\begin{aligned}\int \frac{\mathrm{d}x}{a^2 + x^2} &= \int \frac{a}{a^2 + a^2t^2}\,\mathrm{d}t \\ &= \int \frac{a}{a^2(1 + t^2)}\,\mathrm{d}t \\ &= \frac{1}{a}\int \frac{\mathrm{d}t}{1 + t^2} \\ &= \frac{1}{a}\tan^{-1} t + c \\ &= \frac{1}{a}\tan^{-1}\left(\frac{x}{a}\right) + c\end{aligned}$$

Hence we have

$$\int \frac{\mathrm{d}x}{\sqrt{a^2 - x^2}} = \sin^{-1}\left(\frac{x}{a}\right) + c \quad \text{and} \quad \int \frac{\mathrm{d}x}{a^2 + x^2} = \frac{1}{a}\tan^{-1}\left(\frac{x}{a}\right) + c$$

Example 1 Evaluate $\displaystyle\int_{1.5}^{3} \frac{\mathrm{d}x}{\sqrt{9 - x^2}}$.

$$\begin{aligned}\int_{1.5}^{3} \frac{\mathrm{d}x}{\sqrt{9 - x^2}} &= \left[\sin^{-1}\frac{x}{3}\right]_{1.5}^{3} \\ &= \sin^{-1}(1) - \sin^{-1}\left(\frac{1}{2}\right) \\ &= \frac{\pi}{2} - \frac{\pi}{6} \\ &= \frac{\pi}{3}\end{aligned}$$

Example 2 $\displaystyle\int_{0}^{1} \frac{\mathrm{d}x}{2 + x^2}$.

$$\begin{aligned}\int_{0}^{1} \frac{\mathrm{d}x}{2 + x^2} &= \left[\frac{1}{\sqrt{2}}\tan^{-1}\left(\frac{x}{\sqrt{2}}\right)\right]_{0}^{1} \\ &= \frac{1}{\sqrt{2}}\tan^{-1}\left(\frac{1}{\sqrt{2}}\right) - \frac{1}{\sqrt{2}}\tan^{-1}(0) \\ &= \frac{1}{\sqrt{2}}.\frac{\pi}{4} - 0 \\ &= \frac{\pi}{4\sqrt{2}}\end{aligned}$$

Example 3 Evaluate $\int \frac{dx}{\sqrt{3 - 4x^2}}$.

$$\int \frac{dx}{\sqrt{3-4x^2}} = \int \frac{dx}{\sqrt{4\left(\frac{3}{4} - x^2\right)}}$$

The denominator of the standard form is $\sqrt{a^2 - x^2}$. The coefficient of the x^2 term is 1 so we take out a factor of 4.

$$= \int \frac{dx}{2\sqrt{\left(\frac{3}{4} - x^2\right)}}$$

$$= \frac{1}{2}\int \frac{dx}{\sqrt{\left(\left(\frac{\sqrt{3}}{2}\right)^2 - x^2\right)}}$$

$a^2 = \frac{3}{4} \Rightarrow a = \frac{\sqrt{3}}{2}$

$$= \frac{1}{2}\sin^{-1}\left(\frac{x}{\frac{\sqrt{3}}{2}}\right) + c = \frac{1}{2}\sin^{-1}\left(\frac{2x}{\sqrt{3}}\right) + c$$

EXERCISE 1A

1 Find the following indefinite integrals.

a $\int \frac{dx}{\sqrt{25 - x^2}}$ **b** $\int \frac{dx}{25 + x^2}$ **c** $\int \frac{dx}{\sqrt{1 - x^2}}$ **d** $\int \frac{dx}{x^2 + 3}$

e $\int \frac{dx}{\sqrt{36 - 9x^2}}$ **f** $\int \frac{dx}{48 + 3x^2}$ **g** $\int \frac{dx}{45 + 5x^2}$ **h** $\int \frac{dx}{\sqrt{144 - 16x^2}}$

2 Find the values of the following definite integrals.

a $\int_0^3 \frac{dx}{\sqrt{9 - x^2}}$ **b** $\int_1^2 \frac{dx}{\sqrt{4 - x^2}}$ **c** $\int_1^3 \frac{dx}{3 + x^2}$ **d** $\int_1^{\sqrt{3}} \frac{dx}{1 + x^2}$

e $\int_0^2 \frac{dx}{24 + 6x^2}$ **f** $\int_0^2 \frac{dx}{\sqrt{36 - 4x^2}}$ **g** $\int_1^2 \frac{dx}{\sqrt{12 - 3x^2}}$ **h** $\int_0^1 \frac{dx}{9 + 3x^2}$

3 If $\int_0^2 \frac{dx}{\sqrt{a^2 - 4x^2}} = \frac{\pi}{12}$, find the value of a.

4 The diagram shows a sketch of the curve $y = \frac{1}{\sqrt{9 - 4x^2}}$.
Show that the shaded area is $\sin^{-1}\left(\frac{2}{3}\right)$.

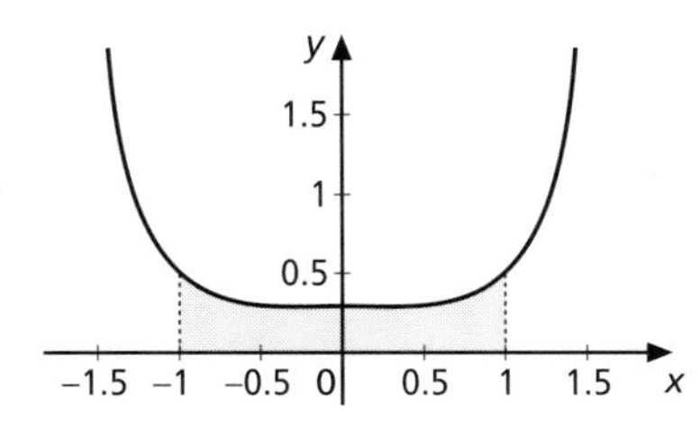

5 If $\int_{\frac{1}{2}}^{a} \frac{dx}{\sqrt{1 - 2x^2}} = \int_0^{\frac{1}{\sqrt{2}}} \frac{dx}{1 + 2x^2}$, find the value of a.

Example Find $\displaystyle\int_3^4 \frac{5}{\sqrt{16 - x^2}}\,dx$.

$$\int_0^2 \frac{5}{\sqrt{16 - x^2}}\,dx = 5\int_0^2 \frac{dx}{\sqrt{16 - x^2}} = 5\left[\sin^{-1}\left(\frac{x}{4}\right)\right]_0^2$$

$$= 5\sin^{-1}\left(\frac{1}{2}\right) - 5\sin^{-1}0$$

$$= \frac{5\pi}{6}$$

EXERCISE 1B

1 Evaluate:

a $\displaystyle\int_0^{\frac{5}{6}} \frac{dx}{\sqrt{25 - 9x^2}}$ **b** $\displaystyle\int_0^1 \frac{dx}{\sqrt{25 - 16x^2}}$ **c** $\displaystyle\int_0^{\frac{1}{\sqrt{2}}} \frac{dx}{3 + 2x^2}$ **d** $\displaystyle\int_0^2 \frac{3}{4 + 3x^2}\,dx$

e $\displaystyle\int_0^1 \frac{dx}{\sqrt{4 - 3x^2}}$ **f** $\displaystyle\int_0^2 \frac{5}{16 + 9x^2}\,dx$ **g** $\displaystyle\int_0^{\frac{1}{2}} \frac{4}{\sqrt{1 - 2x^2}}\,dx$ **h** $\displaystyle\int_0^1 \frac{7}{2x^2 + 1}\,dx$

2 **a** By considering $\displaystyle\int \frac{-1}{\sqrt{a^2 - x^2}}\,dx$ as $\displaystyle -\int \frac{dx}{\sqrt{a^2 - x^2}}$ find the integral $\displaystyle\int \frac{-1}{\sqrt{a^2 - x^2}}\,dx$.

b Find the derivative of $\cos^{-1}\left(\frac{x}{a}\right)$ and hence write down $\displaystyle\int \frac{-1}{\sqrt{a^2 - x^2}}\,dx$.

c The above results indicate that $-\sin^{-1}\left(\frac{x}{a}\right) + C_1 = \cos^{-1}\left(\frac{x}{a}\right) + C_2$ where C_1 and C_2 are constants of integration. If this is rewritten as $\cos^{-1}\left(\frac{x}{a}\right) + \sin^{-1}\left(\frac{x}{a}\right) = C$ where C is a constant, find the value of C.

3 **a** Show that the derivative of $\ln\left|x + \sqrt{x^2 + k}\right|$ is $\dfrac{1}{\sqrt{x^2 + k}}$.

b Hence write down the integral $\displaystyle\int \frac{dx}{\sqrt{x^2 + k}}$.

c Use this result to evaluate, where possible:

(i) $\displaystyle\int_0^2 \frac{dx}{\sqrt{x^2 + 3}}$ **(ii)** $\displaystyle\int_{-1}^1 \frac{dx}{\sqrt{x^2 + 1}}$ **(iii)** $\displaystyle\int_0^1 \frac{dx}{\sqrt{x^2 - 20}}$

(iv) $\displaystyle\int_0^1 \frac{dx}{\sqrt{4x^2 + 16}}$ **(v)** $\displaystyle\int_0^3 \frac{dx}{\sqrt{9x^2 - 27}}$ **(vi)** $\displaystyle\int_{-1}^2 \frac{5}{\sqrt{8x^2 - 12}}\,dx$

Integrals of rational functions

Reminders

- A rational function is one expressed in fractional form whose numerator and denominator are polynomials.
- A rational function is termed *proper* when the degree of the numerator is less than the degree of the denominator. It is termed *improper* otherwise.
- Improper rational functions can be simplified by algebraic division.
- A proper rational function can be resolved into partial fractions.
  There are three basic cases.
  (i) The factors of the denominator are distinct linear functions.
  $$\frac{ax+b}{(cx+d)(ex+f)} = \frac{A}{(cx+d)} + \frac{B}{(ex+f)}$$
  (ii) The denominator contains a repeated linear factor.
  $$\frac{ax^2+bx+c}{(dx+e)(fx+g)^2} = \frac{A}{(dx+e)} + \frac{B}{(fx+g)} + \frac{C}{(fx+g)^2}$$
  (iii) The denominator contains an irreducible quadratic factor.
  $$\frac{ax^2+bx+c}{(dx+e)(fx^2+gx+h)} = \frac{A}{(dx+e)} + \frac{Bx+C}{(fx^2+gx+h)}$$
- In particular examples the values of the upper case constants can be ascertained by suitable selection of convenient values of x.
- $\displaystyle\int \frac{dx}{a+bx} = \frac{1}{b}\ln|a+bx|$
- $\displaystyle\int \frac{f'(x)}{f(x)}\,dx = \ln|f(x)| + c$

Example 1 Find $\displaystyle\int \frac{dx}{x^2-3x+2}$.

$$\int \frac{dx}{x^2-3x+2} = \int \frac{dx}{(x-2)(x-1)} = \int\left(\frac{1}{x-2} - \frac{1}{x-1}\right)dx \qquad \text{partial fractions}$$
$$= \ln|x-2| - \ln|x-1| + c$$
$$= \ln|x-2| - \ln|x-1| + \ln k \qquad \text{letting } c = \ln k \text{ for convenience}$$
$$= \ln\left|\frac{k(x-2)}{x-1}\right| \qquad \text{by the laws of logarithms}$$

Example 2 Find $\displaystyle\int \frac{x}{x^2-6x+9}\,dx$.

$$\int \frac{x}{x^2-6x+9}\,dx = \int \frac{x}{(x-3)^2}\,dx = \int\left(\frac{1}{(x-3)} + \frac{3}{(x-3)^2}\right)dx \qquad \text{partial fractions}$$
$$= \ln|x-3| - \frac{3}{x-3} + c$$

> *Example 3* Find $\int \frac{dx}{x^3 + 4x^2 + x - 6}$.
>
> $$\int \frac{dx}{x^3 + 4x^2 + x - 6} = \int \frac{dx}{(x-1)(x+2)(x+3)} = \int \left(\frac{\frac{1}{12}}{x-1} + \frac{-\frac{1}{3}}{x+2} + \frac{\frac{1}{4}}{x+3} \right) dx$$
>
> $$= \frac{1}{12}\int \frac{dx}{x-1} - \frac{1}{3}\int \frac{dx}{x+2} + \frac{1}{4}\int \frac{dx}{x+3}$$
>
> $$= \frac{1}{12}\ln|x-1| - \frac{1}{3}\ln|x+2| + \frac{1}{4}\ln|x+3| + c$$
>
> $$= \ln\left| \frac{k(x-1)^{\frac{1}{12}}(x+3)^{\frac{1}{4}}}{(x+2)^{\frac{1}{3}}} \right| \quad \text{where } \ln k = c$$

EXERCISE 2

1 Express $\frac{x}{x^2 - 3x + 2}$ in the form $\frac{A}{x-1} + \frac{B}{x-2}$ and hence find $\int \frac{x}{x^2 - 3x + 2} dx$.

2 Express $\frac{x}{x^2 - 4x + 4}$ in the form $\frac{A}{x-2} + \frac{B}{(x-2)^2}$ and hence find $\int \frac{x}{x^2 - 4x + 4} dx$.

3 Express $\frac{2}{x(x+1)(x-1)}$ in the form $\frac{A}{x} + \frac{B}{x+1} + \frac{C}{x-1}$ and hence find $\int_2^3 \frac{2}{x(x+1)(x-1)} dx$.

4 Evaluate $\int_{-7}^{7} \frac{dx}{x^2 + 16x + 64}$.

5 Use partial fractions to help you find each of the following integrals.

a $\int \frac{5x - 11}{2x^2 + x - 6} dx$ **b** $\int \frac{dx}{x^2 - 6x + 9}$ **c** $\int \frac{2}{(x-1)(x-2)(x-3)} dx$

d $\int \frac{10}{x^2 + x - 6} dx$ **e** $\int \frac{2x - 1}{(x+1)(x-1)(x-2)} dx$ **f** $\int \frac{x}{(x+1)^2} dx$

6 Evaluate:

a $\int_1^2 \frac{x + 3}{x^2 - 12x + 27} dx$ **b** $\int_3^4 \frac{12}{(x+2)(x-2)(x-1)} dx$ **c** $\int_0^1 \frac{dx}{x^2 - 10x + 25}$

d $\int_{-1}^{1} \frac{dx}{4x^2 - 12x + 9}$ **e** $\int_2^3 \frac{3x - 2}{(x-1)^2} dx$ **f** $\int_4^5 \frac{6x - 6}{x(x-3)(x+1)} dx$

Example 1 Find $\displaystyle\int \frac{9}{x^3 + 3x^2 - 4}\,dx$.

$$\int \frac{9}{x^3 + 3x^2 - 4}\,dx = \int \frac{dx}{x - 1} - \int \frac{dx}{x + 2} - 3\int \frac{dx}{(x + 2)^2}$$ partial fractions

$$= \ln|x - 1| - \ln|x + 2| + 3\left(\frac{1}{x + 2}\right) + c$$

$$= \ln\left|\frac{x - 1}{x + 2}\right| + \frac{3}{x + 2} + c$$

The constant of integration can be expressed as either c or $\ln k$ here (no further simplification is possible).

Example 2 Find $\displaystyle\int \frac{x}{x^2 + 4}\,dx$.

$$\int \frac{x}{x^2 + 4}\,dx = \frac{1}{2}\int \frac{2x}{x^2 + 4}\,dx$$ convenient manipulation to make the integral of the form $\displaystyle\int \frac{f'(x)}{f(x)}\,dx$

$$= \frac{1}{2}\ln\left|x^2 + 4\right| + c$$

Example 3 Find $\displaystyle\int \frac{x + 3}{x^2 + 4}\,dx$.

$$\int \frac{x + 3}{x^2 + 4}\,dx = \int \frac{x}{x^2 + 4}\,dx + \int \frac{3}{x^2 + 4}\,dx$$

$$= \frac{1}{2}\int \frac{2x}{x^2 + 4}\,dx + 3\int \frac{dx}{x^2 + 4}$$

$$= \frac{1}{2}\ln\left|x^2 + 4\right| + \frac{3}{2}\tan^{-1}\left(\frac{x}{2}\right) + c$$

Example 4 Find $\displaystyle\int \frac{dx}{x^3 - x^2 + 4x - 4}$.

$$\int \frac{dx}{x^3 - x^2 + 4x - 4} = \int \frac{dx}{(x - 1)(x^2 + 4)} = \frac{1}{5}\int \frac{dx}{x - 1} - \frac{1}{5}\int \frac{x + 1}{x^2 + 4}\,dx$$ partial fractions

$$= \frac{1}{5}\int \frac{dx}{x - 1} - \frac{1}{10}\int \frac{2x}{x^2 + 4}\,dx - \frac{1}{5}\int \frac{dx}{x^2 + 4}$$

$$= \frac{1}{5}\ln\left|x - 1\right| - \frac{1}{10}\ln\left|x^2 + 4\right| - \frac{1}{10}\tan^{-1}\left(\frac{x}{2}\right) + c$$

EXERCISE 3A

1 Express $\dfrac{25}{(x-2)(x+3)^2}$ in the form $\dfrac{A}{x-2}+\dfrac{B}{x+3}+\dfrac{C}{(x+3)^2}$ and hence find $\displaystyle\int\frac{25}{(x-2)(x+3)^2}\,dx$.

2 **a** Express $\dfrac{5}{(x+1)(x^2+4)}$ in the form $\dfrac{A}{x+1}+\dfrac{Bx+C}{x^2+4}$.

b Show that your answer is equivalent to $\dfrac{1}{x+1}-\dfrac{1}{2}\left(\dfrac{2x}{x^2+4}\right)+\dfrac{1}{x^2+4}$ and hence find $\displaystyle\int\frac{5}{(x+1)(x^2+4)}\,dx$.

3 Evaluate the following integrals.

a $\displaystyle\int_0^1\frac{dx}{(x+1)(x+2)^2}$

b $\displaystyle\int_2^3\frac{2}{(x-1)(x^2+1)}\,dx$

c $\displaystyle\int_{-1}^1\frac{5}{(x+2)(x^2+1)}\,dx$

d $\displaystyle\int_2^3\frac{8x}{(x-1)(x+1)^2}\,dx$

e $\displaystyle\int_2^3\frac{4x^2}{(x+1)(x-1)^2}\,dx$

f $\displaystyle\int_1^2\frac{2x^2}{(x+1)(x^2+1)}\,dx$

4 Evaluate the following integrals.

a $\displaystyle\int\frac{6x}{(x+1)(x-1)(x-2)}\,dx$

b $\displaystyle\int\frac{4}{x(x-2)^2}\,dx$

c $\displaystyle\int_2^{2.5}\frac{x^2-2}{x(x^2+2)}\,dx$

5 Show that $\displaystyle\int_0^1\frac{8x+14}{(x+1)(x+3)}\,dx=\ln\left(\frac{2^{13}}{3^5}\right)$.

6 **a** By expressing the left-hand side as a single fraction, show that

$$x^2+2x+4+\frac{8}{x-2}=\frac{x^3}{x-2}.$$

b Find the integral $\displaystyle\int\frac{x^3}{x-2}\,dx$.

7 Show that $\displaystyle\int_{2.5}^3\frac{dx}{(x-1)^2(x-2)^2}=\ln\left(\frac{4}{9}\right)+\frac{7}{6}$.

Example Find $\int \frac{x^3 + 2x^2 + 3x + 1}{x^2 + 3x + 2}\,dx$.

Note that we have an improper rational function. This must be simplified by algebraic division and partial fractions.

$$\begin{array}{r}
x - 1 \\
x^2 + 3x + 2 \overline{) x^3 + 2x^2 + 3x + 1} \\
\underline{x^3 + 3x^2 + 2x} \\
-x^2 + x + 1 \\
\underline{-x^2 - 3x - 2} \\
4x + 3
\end{array}$$

$$\frac{x^3 + 2x^2 + 3x + 1}{x^2 + 3x + 2} = x - 1 + \frac{4x + 3}{x^2 + 3x + 2} = x - 1 + \frac{5}{x + 2} - \frac{1}{x + 1}$$

$$\int \frac{x^3 + 2x^2 + 3x + 1}{x^2 + 3x + 2} = \int (x - 1)\,dx + \int \frac{5}{x + 2}\,dx - \int \frac{dx}{x + 1}$$

$$= \frac{x^2}{2} - x + 5\ln|x + 2| - \ln|x + 1| + c$$

EXERCISE 3B

1 Find:

a $\int \frac{2x^3 + 3x^2 - 3}{x^2 - 1}\,dx$

b $\int \frac{3x^4 + 3x^3 + 7x^2 + 7x + 3}{x^3 + x^2 + 2x + 2}\,dx$

c $\int \frac{x^4 + 4x^3 - 9x^2 - 25x - 17}{(x - 3)(x + 1)^2}\,dx$

d $\int \frac{6x^3 - 5x^2 + 19x - 16}{(x - 1)(x^2 + 3)}\,dx$

e $\int \frac{x^4 + 5x^3 + 6x^2 + x - 1}{(x - 1)(x + 2)(x + 3)}\,dx$

f $\int \frac{x^4}{(x^2 + 2x + 1)}\,dx$

2 Show that $\int_1^3 \frac{4}{(x + 1)(x^2 + 3)}\,dx = \frac{1}{2}\ln\left(\frac{4}{3}\right) + \frac{\pi\sqrt{3}}{18}$.

3 Show that $\int_0^1 \frac{4}{(x + 1)(x^2 + 2x + 5)}\,dx = \ln\left(\sqrt{\frac{5}{2}}\right)$.

4 The integral $\int_{\frac{a}{2}}^{a} \frac{dx}{\sqrt{a^2 - x^2}}$ is defined as $\lim_{p \to a} \int_{\frac{a}{2}}^{p} \frac{dx}{\sqrt{a^2 - x^2}}$ provided the limit exists.

Evaluate the integral $\int_{\frac{a}{2}}^{p} \frac{dx}{\sqrt{a^2 - x^2}}$ and hence find $\int_{\frac{a}{2}}^{a} \frac{dx}{\sqrt{a^2 - x^2}}$.

5 The graph shows part of the function

$$f(x) = \frac{x^4 + 1}{x^2 + 3x + 2}.$$

Calculate the shaded area.

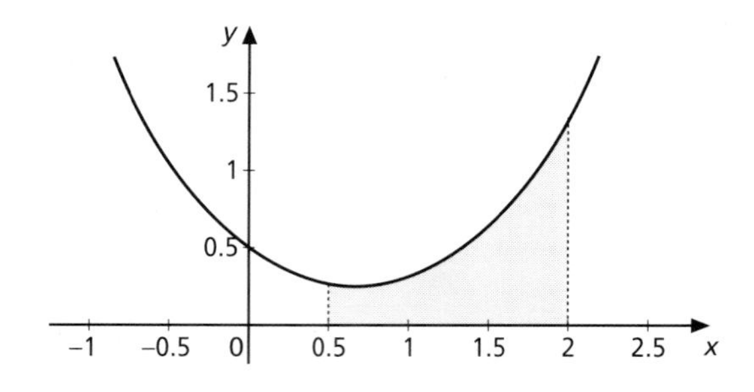

Integration by parts

Sometimes we have to integrate the product of two functions. To see how to do this, let us examine the product rule for differentiation:

$$\frac{d}{dx}(f(x)g(x)) = f'(x)g(x) + f(x)g'(x)$$

Integrating both sides with respect to x:

$$f(x)g(x) = \int f'(x)g(x)\,dx + \int f(x)g'(x)\,dx$$

Rearranging we get:

$g(x)$ differentiated

$$\int f'(x)g(x)\,dx = f(x)g(x) - \int f(x)g'(x)\,dx$$

$f'(x)$ integrated

This formula can be used to exchange the integral of one product $(f'(x)g(x))$ for an expression involving the integral of another $(f(x)g'(x))$. The aim is to make the new integral simpler than the first.

Example 1 Find $\int x \cos x\,dx$.

Note that $\frac{d(x)}{dx} = 1$ whereas $\frac{d(\cos x)}{dx} = -\sin x$. We generally want to choose as $g(x)$ the function which simplifies better on differentiation.

This suggests we use $f'(x) = \cos x$ and $g(x) = x$.

$g(x)$ differentiated

$$\int x \cos x\,dx = x \sin x - \int \sin x.1\,dx$$

$f'(x)$ integrated

$$= x \sin x - \int \sin x\,dx$$

$$= x \sin x + \cos x + c$$

Example 2 Find $\int (3x+2)e^x \, dx$.

Note that $\frac{d(3x+2)}{dx} = 3$ whereas $\frac{d(e^x)}{dx} = e^x$. This suggests we use $f'(x) = e^x$ and $g(x) = 3x + 2$.

$g(x)$ differentiated

$$\int (3x+2)e^x \, dx = (3x+2)e^x - \int e^x.3 \, dx$$

$f'(x)$ integrated

$$= (3x+2)e^x - 3\int e^x \, dx$$
$$= (3x+2)e^x - 3e^x + c$$

EXERCISE 4

1 Use the technique of 'integration by parts' to find the following integrals.

a $\int x \cos x \, dx$ **b** $\int x \sec^2 x \, dx$ **c** $\int (x+2) \ln x \, dx$

d $\int \sin x \ln(\cos x) \, dx$ **e** $\int \frac{\ln x}{x^3} \, dx$ **f** $\int x\sqrt{1+3x} \, dx$

2 Evaluate the following integrals.

a $\int_1^2 x^3 \ln x \, dx$ **b** $\int_0^{\pi} x \cos 3x \, dx$ **c** $\int_0^1 x\sqrt[3]{2-x} \, dx$

d $\int_0^{\frac{\pi}{12}} x \sin 2x \, dx$ **e** $\int_0^1 xe^{2x} \, dx$ **f** $\int_0^1 (2x-3)e^x \, dx$

3 By expressing $\ln x$ as $1 \times \ln x$, find the integral $\int \ln x \, dx$.

4 a Show that $1 - \frac{1}{1+x^2} = \frac{x^2}{1+x^2}$.

b Hence find the integral $\int x \tan^{-1} x \, dx$.

5 Bearing in mind that $\cos 2x = 1 - 2\sin^2 x$, evaluate the integral $\int_0^{\pi} x \sin^2 x \, dx$.

6 Let $P = \int e^x \cos x \, dx$ and $Q = \int e^x \sin x \, dx$.

a For P, use integration by parts once and hence obtain an expression for $P + Q$ which does not involve an integral.

b For Q, use integration by parts once and hence obtain an expression for $P - Q$ which does not involve an integral.

c Hence find expressions for P and Q which do not involve integrals.

Integration by parts: a development

Sometimes the process of integrating by parts must be applied more than once.

Example 1 Find $\int x^2 \sin x \, dx$.

$$\int x^2 \sin x \, dx = x^2(-\cos x) - \int (-\cos x).2x \, dx \qquad \text{first application}$$
$$= -x^2 \cos x + 2\int x \cos x \, dx$$
$$= -x^2 \cos x + 2x \sin x - \int \sin x.2 \, dx \qquad \text{second application}$$
$$= -x^2 \cos x + 2x \sin x + 2 \cos x + c$$

Example 2 Find $\int_0^1 (x^2 + 4x) \cos x \, dx$.

$$\int_0^1 (x^2 + 4x) \cos x \, dx = \left[(x^2 + 4x)(\sin x) - \int (2x + 4) \sin x \, dx\right]_0^1$$
$$= \left[(x^2 + 4x) \sin x - (2x + 4)(-\cos x) + \int 2(-\cos x) \, dx\right]_0^1$$
$$= \left[(x^2 + 4x) \sin x + (2x + 4) \cos x - 2 \sin x\right]_0^1$$
$$= (5 \sin 1 + 6 \cos 1 - 2 \sin 1) - (0 + 4 \cos 0 - 2 \sin 0)$$
$$= 3 \sin 1 + 6 \cos 1 - 4$$

Sometimes the process becomes cyclical.

Example 3 Find $\int e^x \sin x \, dx$.

Note that the derivative of neither function is simpler than the function. Let $f(x) = e^x$.

$$\int e^x \sin x \, dx = e^x \sin x - \int e^x \cos x \, dx$$
$$\int e^x \sin x \, dx = e^x \sin x - e^x \cos x - \int e^x \sin x \, dx$$

The original integral has returned on the right-hand side.

Adding $\int e^x \sin x \, dx$ to both sides gives

$$2\int e^x \sin x \, dx = e^x \sin x - e^x \cos x$$

Thus

$$\int e^x \sin x \, dx = \tfrac{1}{2}(e^x \sin x - e^x \cos x)$$

Note
When we choose to integrate e^x in the first application of the rule, we are then committed to integrate e^x in the second application.

EXERCISE 5A

1 Use the technique of 'integration by parts' to find the following integrals.

a $\int x^2 \cos 2x\,dx$ **b** $\int (x+1)^2 e^{-x}\,dx$ **c** $\int (x^2+3)\sin x\,dx$

d $\int x^3 e^x\,dx$ **e** $\int e^x \cos x\,dx$ **f** $\int e^x \sin 2x\,dx$

g $\int x \ln x\,dx$ **h** $\int x\sqrt{x-1}\,dx$ **i** $\int x(x+4)^{10}\,dx$

j $\int \frac{x}{(x+1)^5}\,dx$ **k** $\int \frac{x^2}{(2x-3)^7}\,dx$ **l** $\int \frac{2x^2}{\sqrt{x+3}}\,dx$

m $\int x \ln x^3\,dx$ **n** $\int x^2 \sin(3x-1)\,dx$ **o** $\int \sin x \sin 5x\,dx$

p $\int \sin 3x \cos 5x\,dx$ **q** $\int \frac{\sin x}{e^x}\,dx$ **r** $\int x3^x\,dx$

s $\int e^{-x}\sin^2 x\,dx$ **t** $\int x\,\mathrm{cosec}^2 x\,dx$ **u** $\int \tan^{-1} x\,dx$

2 Evaluate the following integrals.

a $\int_0^{\frac{\pi}{4}} x^2 \sin 2x\,dx$ **b** $\int_0^{\frac{\pi}{2}} x^2 \cos^2 x\,dx$ **c** $\int_0^1 x^2 e^x\,dx$

d $\int_0^{\frac{\pi}{4}} x \sec x \tan x\,dx$ **e** $\int_1^4 \sqrt{x}\ln x\,dx$ **f** $\int_0^1 \cos^{-1} x\,dx$

g $\int_{\frac{\pi}{6}}^{\frac{\pi}{4}} x\,\mathrm{cosec}^2 x\,dx$ **h** $\int_0^{\frac{\pi}{4}} \sec^3 x\,dx$ **i** $\int_0^{\frac{\pi}{3}} \sin x \ln(\cos x)\,dx$

j $\int_0^1 \frac{x}{(x+1)^4}\,dx$ **k** $\int_3^4 \frac{2x^2}{(x-2)^6}\,dx$ **l** $\int_2^5 \frac{x^2}{\sqrt{x-1}}\,dx$

m $\int_1^e \ln x^4\,dx$ **n** $\int_0^{\frac{\pi}{4}} x^2 \cos\left(4x+\frac{\pi}{4}\right)dx$ **o** $\int_{-1}^1 \cos x \sin 3x\,dx$

p $\int_0^1 \frac{\cos x}{e^{2x}}\,dx$ **q** $\int_0^1 x^2 2^x\,dx$ **r** $\int_{-\pi}^{\pi} e^{-x}\cos 2x\,dx$

3 The diagram shows part of the graph $y = \frac{1}{2}x^2 \sin x$.

Evaluate the area between the curve and the x-axis for $0 \le x \le \pi$.

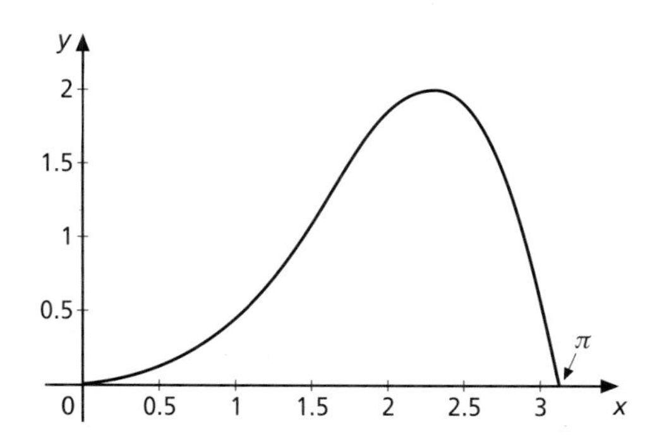

EXERCISE 5B

1 **a** Show that $\frac{1}{2}\left(x - 1\right) + \dfrac{1}{2(1 + x)} = \dfrac{x^2}{2(1 + x)}$.

b Hence find the integral $\int x \ln(1 + x)\,dx$.

2 **a** Show that $1 - \dfrac{4}{4 - 3x} = \dfrac{-3x}{4 - 3x}$.

b Hence find the integral $\int \ln(4 - 3x)\,dx$.

3 Evaluate:

a $\displaystyle\int_1^e (\ln x)^2\,dx$ **b** $\displaystyle\int_0^1 x^3 e^{-2x}\,dx$

4 Show that $2 - \dfrac{2}{1 + x^2} = \dfrac{2x^2}{1 + x^2}$ and hence evaluate $\displaystyle\int_0^{\sqrt{3}} \ln(1 + x^2)\,dx$.

5 By treating $\displaystyle\int_0^{\infty} e^{-\frac{x}{3}} \sin 2x\,dx$ as $\displaystyle\lim_{p\to\infty}\int_0^{p} e^{-\frac{x}{3}} \sin 2x\,dx$ find the value of $\displaystyle\int_0^{\infty} e^{-\frac{x}{3}} \sin 2x\,dx$.

6 Use the substitution $x = t^2$ to help you find the integral $\int \cos\sqrt{x}\,dx$.

7 When the curve $y = x(1 + x)^{\frac{3}{4}}$ is rotated about the x-axis in the domain $-1 \le x \le 2$ the solid shown here is generated.

The volume of such a solid can be calculated using the formula

$$V = \pi\int_a^b y^2\,dx$$

where the curve is rotated in the domain

$$a \le x \le b$$

Calculate the volume of the solid illustrated.

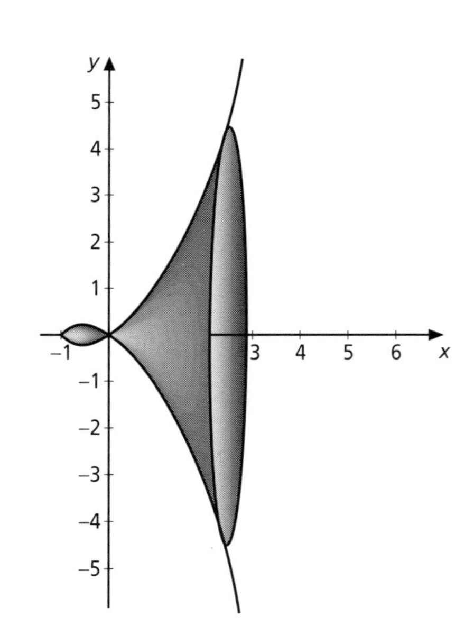

Differential equations

If an equation contains a derivative then it is called a *differential equation.*
A function which satisfies the equation is called a *solution* of the differential equation.

The *order* of a differential equation is the order of the highest derivative involved.
The *degree* of a differential equation is the degree of the power of the highest derivative involved.

For example

$\frac{dy}{dx} = 2x - 4$	1st order	1st degree
$(1 - x^2)\frac{dy}{dx} - 3y = 4$	1st order	1st degree
$\left(\frac{dy}{dx}\right)^2 + x + 4y = 0$	1st order	2nd degree
$\frac{d^2y}{dx^2} + 4\frac{dy}{dx} + 3x = 0$	2nd order	1st degree

This chapter deals only with first-order, first-degree differential equations, and even then only with equations where the two variables involved can be separated from each other on opposite sides of the equation.

Differential equations are solved by integration. When the solution contains the constant of integration it is called a *general solution.*

When we are given some *initial conditions* which allow us to evaluate this constant the resultant solution is called a *particular solution.*

Example 1

a Find the general solution of the differential equation $\frac{dy}{dx} = x^3 + 4x$.

b Find the particular solution corresponding to the initial conditions that $y = 3$ when $x = 2$.

a Integrating with respect to x:

$$y = \int x^3 + 4x\,dx$$
$$= \frac{1}{4}x^4 + 2x^2 + c$$

This is the general solution (c is an arbitrary constant).

b When $x = 2$, $y = 3$, the equation becomes

$$3 = 4 + 8 + c$$
$$\Rightarrow \quad c = -9$$
$$\Rightarrow \quad y = \frac{1}{4}x^4 + 2x^2 - 9$$

This is the particular solution associated with the initial conditions $x = 2$, $y = 3$.

Example 2 Find the particular solution to $\frac{dy}{dx} = x^2 + e^x$ given that, when $x = 0$, $y = 0$.

$$\frac{dy}{dx} = x^2 + e^x$$

$$\Rightarrow \quad y = \int x^2 + e^x \, dx$$

$$\Rightarrow \quad y = \frac{1}{3}x^3 + e^x + c \qquad \text{general solution}$$

$$\Rightarrow \quad 0 = 0 + 1 + c \qquad \text{using initial condition}$$

$$\Rightarrow \quad c = -1$$

$$\Rightarrow \quad y = \frac{1}{3}x^3 + e^x - 1 \qquad \text{particular solution}$$

EXERCISE 6

1 Find the general solution of each of the following differential equations.

a $\frac{dy}{dx} = 2x$ **b** $\frac{dy}{dx} = ax$ **c** $\frac{dy}{dx} = x^2 - 1$

d $\frac{dy}{dx} = (x - 2)(x + 1)$ **e** $\frac{dy}{dx} = \sec^2 3x$ **f** $\frac{dy}{dx} = \frac{1}{\sqrt{a^2 - x^2}}$

g $\frac{dy}{dx} = \frac{1}{3x + 2}$ **h** $\frac{dy}{dx} = \cos^2 x$ **i** $\frac{dy}{dx} = \cot 4x$

j $\frac{dy}{dx} = \frac{1}{(3x - 7)^5}$ **k** $\frac{dy}{dx} = \frac{1}{16 + x^2}$ **l** $\frac{dy}{dx} = \frac{1}{16 + 4x^2}$

2 Find the particular solution of each of the following differential equations.

a $\frac{dy}{dx} = x^3$ given that $y = 1$ when $x = 2$

b $\frac{dy}{dx} = \sin nx$ given that $y = 5$ when $x = \pi$

c $\frac{dy}{dx} = \frac{1}{4 + x^2}$ given that $y = \frac{\pi}{6}$ when $x = 2$

d $\frac{dy}{dx} = \frac{2}{x - 1} - 3x$ given that $y = 0$ when $x = 2$

e $\frac{dy}{dx} = e^{3x+2}$ given that $y = e^2$ when $x = 0$

f $\frac{dy}{dx} = \sec^2 4x$ given that $y = 1$ when $x = \frac{\pi}{16}$

3 Using partial fractions, find the particular solution to the equation $\frac{dy}{dx} = \frac{1}{x(x - 1)}$ given that $y = 0$ when $x = 2$.

4 Using integration by parts, find the particular solution to the equation $\frac{dy}{dx} = xe^{2x}$ given that $y = 0$ when $x = 1$.

5 Show that, for the equation $\frac{dy}{dx} = \frac{2}{x^2 - 1}$, if $y = \ln 2$ when $x = 2$ then $y = \ln\left|\frac{2(x-1)}{x+1}\right|$.

6 The equation $\frac{dy}{dx} = 4x\cos 2x$ has a particular solution when $x = \frac{\pi}{2}$ and $y = 0$.
Show that $y = 2x\sin 2x + \cos 2x + 1$.

Further differential equations

When a differential equation is of the form $\frac{dy}{dx} = f(y)$ then we can use differentials and separate the variables:

$$\frac{1}{f(y)}\,dy = dx$$

We can then integrate both sides

$$\int \frac{dy}{f(y)} = \int dx$$

to obtain a general solution.

Example 1 Find the general solution to the equation $\frac{dy}{dx} = y$.

$$\frac{dy}{dx} = y$$
$$\Rightarrow \quad \frac{1}{y}dy = dx$$
$$\Rightarrow \quad \int \frac{1}{y}dy = \int dx$$
$$\Rightarrow \quad \ln|y| = x + c$$
$$\Rightarrow \quad y = e^{x+c}$$

Alternatively we can write $y = ke^x$, where $k = e^c$, since $e^{x+c} = e^x e^c$.

Example 2 Find the general solution to the equation $\frac{dy}{dx} = \frac{3}{\sqrt{3y+1}}$.

$$\sqrt{3y+1}\,dy = 3\,dx$$
$$\Rightarrow \quad \int \sqrt{3y+1}\,dy = \int 3\,dx$$
$$\Rightarrow \quad \frac{2}{9}(3y+1)^{\frac{3}{2}} = 3x + c$$
$$\Rightarrow \quad y = \frac{1}{3}\left(\frac{9}{2}(3x+c)\right)^{\frac{2}{3}} - 1$$

Example 3 Find the particular solution to the equation $\frac{dy}{dx} = y(y+1)$ if $y = 1$ when $x = 0$.

$$\frac{dy}{dx} = y(y+1)$$

$$\Rightarrow \quad \frac{dy}{y(y+1)} = dx \quad \Rightarrow \quad \int \frac{dy}{y(y+1)} = \int dx$$

$$\Rightarrow \quad \int \frac{dy}{y} - \int \frac{dy}{y+1} = \int dx \qquad \text{partial fractions are required here}$$

$$\Rightarrow \ln|y| - \ln|y+1| = x + c$$

$$\Rightarrow \quad \ln\left|\frac{y}{y+1}\right| = x + c \qquad \text{using the laws of logs}$$

$$\Rightarrow \quad \frac{y}{y+1} = e^{x+c} = e^x e^c = ke^x \qquad \text{where } k = e^c$$

$$\Rightarrow \quad y = ke^x y + ke^x$$

$$\Rightarrow \quad y(1 - ke^x) = ke^x \qquad \text{rearranging}$$

$$\Rightarrow \quad y = \frac{ke^x}{1 - ke^x} \qquad \text{the general solution}$$

Using the initial conditions

$$1 = \frac{k}{1-k} \quad \Rightarrow \quad k = \tfrac{1}{2} \quad \Rightarrow \quad y = \frac{\frac{1}{2}e^x}{1 - \frac{1}{2}e^x}$$

$$\Rightarrow \quad y = \frac{e^x}{2 - e^x} \qquad \text{the particular solution}$$

EXERCISE 7

1 Find the general solution of each of the following differential equations.

a $\frac{dy}{dx} = 2y$ **b** $\frac{dy}{dx} = y^3$ **c** $\frac{dy}{dx} = ay$ **d** $\frac{dy}{dx} = y^2 - 1$

e $\frac{dy}{dx} = -\sqrt{9 - y^2}$ **f** $\frac{dy}{dx} = 3 + y^2$ **g** $\frac{dy}{dx} = e^{3y}$ **h** $\frac{dy}{dx} = y^2 - 3y + 2$

2 Find the particular solution of each of the following differential equations.

a $\frac{dy}{dx} = 3 - y$ given that $y = 0$ when $x = 0$

b $\frac{dy}{dx} = y^2 - y - 2$ given that $y = 3$ when $x = 0$

c $\frac{dy}{dx} = \sqrt{y^2 - 1}$ given that $y = 1$ when $x = 1$

d $\frac{dy}{dx} = \cos^2 y$ given that $y = \frac{\pi}{4}$ when $x = 1$

e $\frac{dy}{dx} = 3 + y^2$ given that $y = 1$ when $x = 0$

f $\frac{dy}{dx} = (2y - 3)^5$ given that $y = 2$ when $x = 1$

Variables separable

When a differential equation can be expressed in the form $\frac{dy}{dx} = f(x)g(y)$ then again we separate the variables and integrate both sides.

$$\frac{1}{g(y)}\,dy = f(x)\,dx \quad \Rightarrow \quad \int \frac{dy}{g(y)} = \int f(x)\,dx$$

Example 1 Find the general solution to the equation $\frac{dy}{dx} = \frac{y^2}{2x+1}$.

$$\frac{dy}{dx} = \frac{y^2}{2x+1}$$

$$\Rightarrow \quad \frac{1}{y^2}\,dy = \frac{1}{2x+1}\,dx \qquad \text{separating the variables}$$

$$\Rightarrow \quad \int \frac{dy}{y^2} = \int \frac{dx}{2x+1}$$

$$\Rightarrow \quad -\frac{1}{y} = \tfrac{1}{2}\ln|2x+1| + \ln k \qquad \text{where } c = \ln k$$

$$\Rightarrow \quad -\frac{1}{y} = \ln\left(k\,|2x+1|^{\frac{1}{2}}\right) \quad \Rightarrow \quad y = -\frac{1}{\ln\left(k\,|2x+1|^{\frac{1}{2}}\right)}$$

Example 2 Find the particular solution to the equation $\frac{dy}{dx} = (3x^2+1)e^y$ given that $y = 0$ when $x = 0$.

$$\frac{dy}{dx} = (3x^2+1)e^y$$

$$\Rightarrow \quad \int \frac{dy}{e^y} = \int (3x^2+1)\,dx$$

$$\Rightarrow \quad -\frac{1}{e^y} = x^3 + x + c \quad \Rightarrow \quad c = -x^3 - x - \frac{1}{e^y}$$

$$x = 0,\ y = 0 \Rightarrow c = -1$$

$$\Rightarrow \quad -\frac{1}{e^y} = x^3 + x - 1 \quad \Rightarrow \quad e^y = \frac{1}{1-x-x^3}, \quad 1 - x - x^3 \neq 0$$

$$\Rightarrow \quad y = \ln\left(\frac{1}{1-x-x^3}\right), \quad 1 - x - x^3 > 0$$

EXERCISE 8

1 Find the general solution of each of the following differential equations.

a $y = x\frac{dy}{dx}$ **b** $3y\frac{dy}{dx} = 4x$ **c** $2y = (x+1)\frac{dy}{dx}$

d $x^2\frac{dy}{dx} + y^2 = 0$ **e** $3(y+2) = 2(x+3)\frac{dy}{dx}$ **f** $\frac{dy}{dx} = \frac{y}{\sqrt{3-x^2}}$

g $y - x\frac{dy}{dx} = 1 + x^2\frac{dy}{dx}$ **h** $\cos^2 y\frac{dy}{dx} + e^x = 0$ **i** $\cos^2 y + e^x\frac{dy}{dx} = 0$

2 Find the particular solution of each of the following differential equations.

a $\frac{dy}{dx} = \frac{3x^2 + 1}{4y + 2}$ given that $y = 1$ when $x = 1$

b $\sec y + e^x \frac{dy}{dx} = 0$ given that $y = \frac{\pi}{6}$ when $x = 0$

c $x\frac{dy}{dx} = (1 + x)y^2$ given that $y = 1$ when $x = 1$

3 **a** Use the technique of integrating by parts to find the integral $\int xe^{-x}\,dx$.

b Hence find the particular solution to the equation $e^x\frac{dy}{dx} = xy^2$ given that $y = 1$ when $x = 0$.

4 **a** Find the integral $\int \frac{\cos t}{\sin t}\,dt$.

b Hence find the particular solution to the equation $\tan x\frac{dy}{dx} = 1 + y^2$ given that $y = 1$ when $x = \frac{\pi}{2}$.

5 **a** Express $\frac{1}{t^2 - 3t + 2}$ in partial fractions.

b Hence find the particular solution to the equation $\frac{dy}{dx} = x(y^2 - 3y + 2)$ given that $y = 3$ when $x = 0$.

6 **a** Find the integral $\int \frac{dt}{t\ln t}$ using the substitution $\ln t = u$.

b Hence find the particular solution to the equation $x\frac{dy}{dx} = y\ln y$ given that $y = e^2$ when $x = 1$.

7 **a** Show by differentiation that $\ln(\operatorname{cosec} x - \cot x)$ is an antiderivative of $\operatorname{cosec} x$.

b Hence find the particular solution to the equation $\sin y + e^x\frac{dy}{dx} = 0$ given that $y = \frac{\pi}{2}$ when $x = 0$.

8 Find the general solution to the equation $\frac{dy}{dx} = x(y^2 + y + 1)$.
[Hint: complete the square.]

9 The equation $x\frac{dy}{dx} + y^2 - 1 = 0$ has a particular solution $y = 2$ when $x = 1$. Show that $y = \frac{3x^2 + 1}{3x^2 - 1}$.

10 If the equation $\frac{dy}{dx} = \frac{1 + y^2}{1 + x^2}$ has a particular solution of $y = 1$ when $x = 0$, show that $y = \frac{1 + x}{1 - x}$.

$\left[\text{You will need the expansion formula } \tan(A + B) = \frac{\tan A + \tan B}{1 - \tan A \tan B}.\right]$

Applications of differential equations

Historical note

Daniel Bernoulli

Like many branches of mathematics, the methods for solving differential equations have been developed as the need arose from physical situations. Using these methods, Daniel Bernoulli was able to develop his theories on fluid flow – *Bernoulli's Principle* is used in the development of aircraft, ships, bridges and other structures where air or water pressure plays a vital part.

The Tacoma Bridge in the USA was nicknamed Galloping Gertie because it oscillated spectacularly in the wind. Clearly the mathematics of the bridge was not correctly computed as it collapsed four months after it was opened.

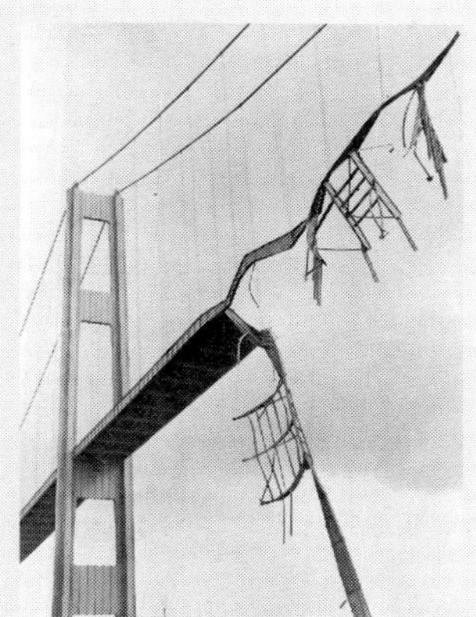
Tacoma Bridge, USA

Example 1 Newton's law of cooling states that the *rate at which an object cools* is proportional to the *difference* between its temperature and that of its surroundings. Let T be the temperature difference at time t. If $T = T_0$ at $t = 0$, express T in terms of t.

$$\frac{\mathrm{d}T}{\mathrm{d}t} \propto T$$

$$\Rightarrow \frac{\mathrm{d}T}{\mathrm{d}t} = kT \qquad k \text{ is the constant of proportion}$$

$$\Rightarrow \frac{1}{T}\mathrm{d}T = k\mathrm{d}t \Rightarrow \int \frac{1}{T}\mathrm{d}T = \int k\mathrm{d}t$$

$$\Rightarrow \ln T = kt + c \qquad c \text{ is the constant of integration}$$

$$\Rightarrow \ln\frac{T}{C} = kt \qquad \ln C = c \text{ for convenience}$$

$$\Rightarrow T = Ce^{kt}$$

If $T = T_0$ when $t = 0$ then $T_0 = C$. Hence $T = T_0e^{kt}$.

Example 2 Given the differential equation $\frac{\mathrm{d}y}{\mathrm{d}x} = -\frac{x}{y}$, find the general solution and illustrate it graphically.

$$\frac{\mathrm{d}y}{\mathrm{d}x} = -\frac{x}{y} \Rightarrow y\,\mathrm{d}y = -x\,\mathrm{d}x$$

$$\Rightarrow \int y\,\mathrm{d}y = -\int x\,\mathrm{d}x \Rightarrow \frac{y^2}{2} = -\frac{x^2}{2} + c$$

$$\Rightarrow \frac{y^2}{2} + \frac{x^2}{2} = c \Rightarrow y^2 + x^2 = 2c$$

$$\Rightarrow y = \pm\sqrt{2c - x^2}$$

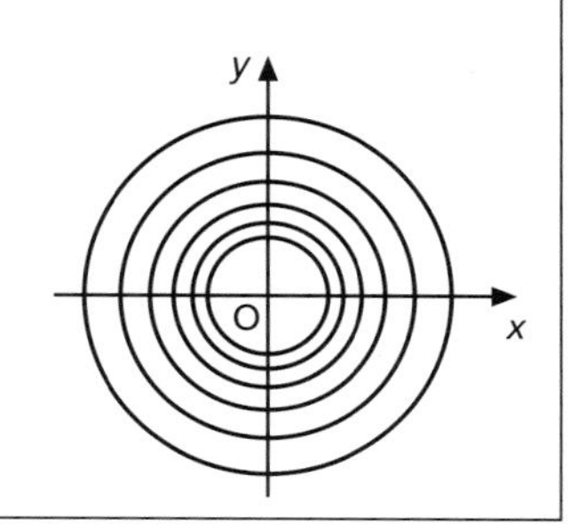

This is a family of circles, centre (0, 0) and radius $\sqrt{2c}$.

Example 3 Two curves are said to be *orthogonal* if, at their point of intersection, the tangent of one is perpendicular to the tangent of the other.
An *orthogonal trajectory* of a family of curves is a curve which is orthogonal to each member of the family. For instance, the line $y = 3x$ is an orthogonal trajectory of the family of circles $x^2 + y^2 = 2c$.
Find the orthogonal trajectory passing through (1, 0) of the family of parabolae $y = x^2 + c$.

$$y = x^2 + c$$
$$\Rightarrow \quad \frac{dy}{dx} = 2x$$

The gradient at the point (x, y) on each parabola is $2x$.
So the gradient at the point (x, y) on the orthogonal trajectory is $-\frac{1}{2x}$ $(x \neq 0)$.

For the orthogonal trajectory

$$\frac{dy}{dx} = -\frac{1}{2x}$$
$$\Rightarrow \int dy = -\int \frac{dx}{2x}$$
$$\Rightarrow \quad y = -\tfrac{1}{2}\ln|x| + c$$

Given $y = 0$ when $x = 1$, we get $c = 0$. Hence

$y = -\frac{1}{2}\ln|x|$ is the required equation.

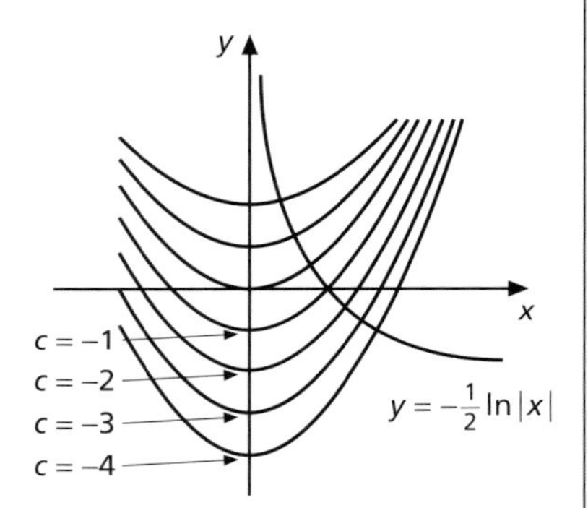

Example 4 Population models come in various guises. One theory states that the rate of change of population is proportional to the population.
If $P = P_0$ at time $t = 0$, express P in terms of t.

$$\frac{dP}{dt} \propto P$$
$$\Rightarrow \quad \frac{dP}{dt} = kP \quad \text{where } k \text{ is a constant}$$
$$\Rightarrow \int \frac{dP}{P} = \int k\,dt$$
$$\Rightarrow \quad \ln P = kt + \ln C$$
$$\Rightarrow \ln \frac{P}{C} = kt$$
$$\Rightarrow \quad P = Ce^{kt}$$

Since $P = P_0$ at time $t = 0$ we have $C = P_0$. Hence

$$P = P_0 e^{kt}$$

EXERCISE 9A

1 Write down a mathematical model in the form of a differential equation for each of these statements.

a The rate of change of displacement (s) with time (t) is directly proportional to the displacement.

b The rate of change of displacement (s) with time (t) is inversely proportional to the time.

c The rate of increase in the number of people (n) who have heard a rumour at a time t is directly proportional to the number who know it at that time.

d A sphere of volume V and radius r is expanding such that the rate of increase in its volume is directly proportional to the size of its equator.

e There are 500 football stickers to collect to fill an album. The rate at which the number of stickers, N, in the album increases is directly proportional to the number of stickers still to collect.

f The rate of change of the number of books bought, B, with respect to cost, is inversely proportional to the cost of a book £C.

2 As you climb a mountain by the sea, your view to the horizon improves. The rate at which the distance to the horizon, D km, changes with respect to the height climbed, h m, can be obtained from the model

$$\frac{\mathrm{d}D}{\mathrm{d}h} = \frac{k}{\sqrt{h}} \quad \text{where } k \text{ is a constant}$$

At sea level ($h = 0$), the distance to the horizon is 8 km.

a Express D in terms of h and constants.

b After climbing 100 metres above sea level the distance to the horizon is 55 km. Find a formula which relates D to h.

c How far can you see to the horizon when you have climbed 850 m?

3 Over the course of an illness, a patient's temperature is measured and averaged, T°F, daily. Its change can be modelled by $\frac{\mathrm{d}T}{\mathrm{d}t} = -\frac{2}{5}t + \frac{3}{2}$ where t is the time measured in days.

a When $t = 0$ at the onset of the illness, the patient's temperature was normal: 98.4 °F. Express T in terms of t explicitly.

b How many days elapse before his temperature returns to normal?

4 A new television series has a prospective target audience in mind. As the weeks go by after the launch of the series the percentage of the target audience watching the programme, V%, is expected to grow according to the model $\frac{\mathrm{d}V}{\mathrm{d}t} = ke^{-0.3t}$ where t is measured in weeks and k is a constant.

a Express V in terms of t and the constants k and c where c is the constant of integration.

b Initially none of the audience are watching ($t = 0$, $V = 0$). Express c in terms of k.

c After five weeks 77.7% of the target audience are watching. Express V in terms of t.

d After 10 weeks 90% of the target audience are watching. Is the model holding up? Comment.

5 When the rate of growth (or decay) of a quantity is proportional to the magnitude of the quantity, then the growth is said to be *unrestricted.*

A rumour spreads through a population in such a way that the rate at which the number of people who have heard the rumour, n, grows is proportional to the number of people who have heard the rumour, i.e. $\frac{\mathrm{d}n}{\mathrm{d}t} = kn$ where k is a constant of proportion. Express n explicitly in terms of t given that, when $t = 0$, $n = 3$ (three people started the rumour) and that, when $t = 2$, $n = 60$.

6 The rate at which the value of a car, £V, depreciates at any given time, is proportional to the value of the car at that time.

a Write down a differential equation involving V and t, where V is the value of the car after t years.

b Initially, $t = 0$, the car is valued at £20 000. After two years the car has decreased in value by £2500. Find the particular solution to the differential equation to express V in terms of t.

c **(i)** When will the car be worth only £10 000?
(ii) What will the car be worth in 10 years?

7 Another growth model is when there is a limit to the growth. The quantity grows inside a finite space. The rate of growth of the quantity at any time is proportional to the space still available in which to grow.

A boy is collecting stickers to put in an album. There is space for 500 different stickers. The boy buys stickers every day. Let t be the number of days since he started his collection. The rate of growth of the number of stickers, N, in his collection at any time t is directly proportional to the number of stickers still to collect, i.e. $\frac{\mathrm{d}N}{\mathrm{d}t} = k(500 - N)$ where k is the constant of proportion.

a Show that the general solution is $N = 500 - e^{-(kt+c)}$ where c is the constant of integration.

b At time $t = 0$ the boy has no stickers ($N = 0$). What is the value of c?

c When $t = 4$ he has 50 stickers. Calculate the value of k.

d When he needs only 50 stickers to complete his collection he can send away for them. After how many days can he do this?

8 A learner driver has to learn 400 facts for her test. The rate at which the number of facts she can recall grows is proportional to the number of facts still to memorise. She applies herself regularly to the task on a daily basis. Her learning can be modelled by $\frac{\mathrm{d}F}{\mathrm{d}t} = k(400 - F)$ where F is the number of facts memorised and t is the time measured in days since she started the task. Initially she knew no facts. After five days she could memorise 250 facts.

a Express F explicitly in terms of t.

b When she can remember 80% of the facts she can claim to have mastery of the topic. For how many days will she have to study to claim mastery?

9 In a third model, the *restricted growth model*, the quantity again has a limit to which it can grow. The rate of increase of the quantity at a particular time is jointly proportional to the quantity present at that time and the quantity needed to reach the limit.

In a small island there are 2000 inhabitants. An islander returns from his holidays and brings a flu virus with him. It spreads among the population according to the model $\frac{dP}{dt} = kP(2000 - P)$ where P is the number with the virus after t days and k is the constant of proportion.

a With the aid of partial fractions show that $\frac{1}{2000}\ln\left(\frac{P}{2000 - P}\right) = kt + c$ where k and c are constants.

b Assuming that $P = 1$ when $t = 0$, find the value of c to one significant figure.

c If 20 people have contracted the virus after five days

(i) express t in terms of P, **(ii)** express P in terms of t.

d Help will have to be flown in if more than 50% of the population contract the virus. Estimate the number of days it will be before this happens.

10 Mildew hits a crop of corn in a field. Its spread can be modelled by $\frac{dP}{dt} = kP(100 - P)$ where P is the percentage of the field affected in day t.

When $t = 0$, $P = 1$. When $t = 5$, $P = 60$.

a Express P in terms of t.

b Estimate the time it will take for 80% of the crop to be affected.

EXERCISE 9B

1 An object is cooling according to Newton's law of cooling, i.e. the rate of change of temperature, $\frac{dT}{dt}$ is proportional to the temperature difference between the object and its surroundings. Let T be the temperature of the object and T_s the temperature of the surroundings, both in °C.

a Write down a differential equation describing this situation.

b When $T = 50$ °C, $\frac{dT}{dt} = -0.16$ °C/s. Determine the constant of proportion.

c Find the rule for T (i.e. find the general solution to the differential equation).

d **(i)** If the temperature of the body is 70 °C at time $t = 0$ and the temperature of the surroundings, T_s, remains constant at 40 °C, find the particular solution.

(ii) Assuming that the temperature of the surrounding air remains constant, calculate how much the temperature of the object has fallen after 1 minute.

2 A cup of cold water is warming up according to Newton's law. Let T be the temperature difference (in °C) between the cup and the surrounding air after t seconds. When T is 5 °C, the rate of warming is –0.005 °C/s.

Assuming that the temperature of the surrounding air remains constant at 12 °C, calculate the rise in temperature of the cup of water by 10.15 a.m. if the temperature difference at 10 a.m. was 9 °C.

3 The gradient of a straight line through the point (x, y) is 4.

a Write down a differential equation describing this situation.

b Find the general solution of this equation and illustrate the result graphically.

4 Sketch the family of curves which represent

a the general solution to the equation $\frac{dy}{dx} = -\frac{x}{y}$

b the general solution to the equation $\frac{dx}{dy} = -\frac{x-a}{y-b}$

5 The number of radioactive atoms in a substance falls as time progresses. The rate of decay, $\frac{dn}{dt}$, of the number of radioactive atoms is proportional to n, the number of atoms present at time t seconds.
In one particular sample there are initially 200 radioactive atoms present. After 10 000 seconds there remain 199 radioactive atoms. The *half-life* of a substance is defined as the time it takes for the number of radioactive atoms to halve.
Determine the half-life of the substance.

6 In physiology the equation $\frac{dh}{dt} = -a\left(\frac{h}{b+h}\right)$ models the hormone concentration, h, in the bloodstream after t seconds where a and b are constants.
Find the general solution of this equation.

7 During the first 30 days of growth for corn and barley, the rate of growth in grams per day is given by the equation $\frac{dW}{dt} = \frac{1}{5}W$ where W is the weight after t days.
If the weight of the plant at the start is 0.06 g, determine its weight after 30 days.

8 The supergrowth model for world population is given by the equation $\frac{dP}{dt} = 0.015P^{1.2}$ where P is the population of the world in millions and t is the time in years with $t = 0$ representing 1990.
Given that the population in 1990 was 4000 million find the model's prediction for the world population in the year 2000.

9 Consider a particular object falling through the air. Its velocity increases due to the force of gravity. If the object is x metres above the Earth at time t seconds and falling with velocity $v\,\text{m s}^{-1}$, Newton's laws of motion give us $\frac{dv}{dt} = -g$ where g is the gravitational constant ($\approx 9.8\,\text{m s}^{-2}$). At the start $t = t_0$, $v = v_0$ and $x = x_0$.

a Determine an equation for v in terms of t.

b By definition of the velocity, $v = \frac{dx}{dt}$. Express x explicitly in terms of t.

10 You borrow £50 000 for a mortgage. The interest rate is 6%.

a In a 'simple-interest bank' they calculate the interest due each month so that the amount owed after one year is given by $V = \left(1 + \frac{0.06}{12}\right)^{12} V_0$ where V_0 is the initial amount borrowed. Calculate how much is owed at the end of one year on a mortgage loan of £50 000.

b In a more realistic bank they calculate the interest *continuously*. The amount owed after one year, V, is modelled by the differential equation $\frac{dV}{dt} = 0.06V$. Calculate the *extra* interest charged at the end of one year if the interest is calculated continuously.

11 In 1845 Otto Verhulst introduced the logistic equation for population models (see the restricted growth model in question 9 of Exercise 9A).

In one particular country the population is modelled by $\frac{dP}{dt} = \frac{P}{625}(200 - P)$ where P is the population measured in millions and t is the time measured in units of 10 years. At time $t = 0$, $P = 4$ million.

a Find, with the aid of partial fractions, the general solution to the differential equation.

b Estimate the population after 10 years.

12 The rate at which petrol leaks from the bottom of a tank is proportional to the square root of the depth of the water above the bottom of the tank.

Thus we can model the situation by $\frac{dV}{dt} = -k\sqrt{V}$ where V is volume of petrol in the tank at time t hours and k is a *positive* constant.

Consider a cuboid tank with a base of area $5\,\text{m}^2$ and height $20\,\text{m}$, initially full. The tank is losing petrol at the rate of $2\,\text{m}^2$ per hour when it is a quarter full.

a Determine a formula modelling the volume of petrol, V, in the tank at time t.

b How long will the tank take to empty?

13 The motion of a body falling due to gravity and under the influence of air resistance can be modelled by $\frac{dv}{dt} = -g - bv$ where g is the gravitational constant ($\approx 9.8\,\text{m}\,\text{s}^{-2}$) and b is a constant (the coefficient of air resistance, ≈ 0.2 in this question).

A body is held 200 m above the ground and released. Find formulae for the velocity, $v\,\text{m}\,\text{s}^{-1}$, and distance, $x\,\text{m}$, above the ground after t seconds.

14 $\frac{dy}{dx} - 3x^2y = x^2$ is an example of a differential equation where we cannot separate the variables. Verify that $y = -\frac{1}{3} + Ce^{x^3}$ is a solution of this equation (C being a constant).

CHAPTER 3 REVIEW

1 Find the values of the following integrals.

a $\displaystyle\int_3^6 \frac{dx}{\sqrt{36 - x^2}}$ **b** $\displaystyle\int_0^2 \frac{dx}{25 + x^2}$ **c** $\displaystyle\int_{\sqrt{3}}^{3\sqrt{3}} \frac{4}{54 + 6x^2}\,dx$

2 Carry out the following integrations.

a $\displaystyle\int \frac{dx}{(x - 3)(x + 2)}$ **b** $\displaystyle\int \frac{dx}{(x + 2)(x^2 + 1)}$

c $\displaystyle\int \frac{4}{(x - 1)(x + 1)(x + 2)}\,dx$ **d** $\displaystyle\int \frac{5x + 2}{(x + 1)(x^2 + x + 5)}\,dx$

e $\displaystyle\int \frac{x^3 + x}{(x + 1)(x - 2)}\,dx$

3 Use the technique of 'integrating by parts' to find the following integrals.

a $\displaystyle\int (4x - 3)e^{2x}\,dx$ **b** $\displaystyle\int_0^{\frac{\pi}{4}} 3x^2 \sin 2x\,dx$

4 Find the general solution to each of the following equations.

a $\displaystyle\frac{dy}{dx} = y(2x + 1)$ **b** $\displaystyle\frac{dy}{dx} = \sec y$

5 Find the particular solution to the equation $\displaystyle\frac{dy}{dx} = 4y^2e^{2x}$ given that $y = -\frac{1}{2}$ when $x = 0$.

6 The rate at which an object cools is directly proportional to the temperature difference between the object and the surrounding air.

If the surrounding temperature is 60 °C we have $\displaystyle\frac{dT}{dt} \propto (T - 60)$ where T is the temperature of the object after t minutes.

An object cools from 100 °C to 80 °C in 30 minutes when the temperature of the surrounding air is 60 °C.

Assuming that the temperature of the surrounding air remains constant, find the temperature of the object at the end of the next 30 minutes.

7 The gradient of the tangent at the point (x, y) on a curve is given by the expression $\displaystyle\frac{-3x}{\sqrt{1 - x^2}}$. Write down the appropriate differential equation and sketch the family of curves which represent the general solution.

CHAPTER 3 SUMMARY

1 *Useful standard integrals*

$$\int \frac{dx}{\sqrt{a^2 - x^2}} = \sin^{-1}\left(\frac{x}{a}\right) + c \qquad \int \frac{dx}{a^2 + x^2} = \frac{1}{a}\tan^{-1}\left(\frac{x}{a}\right) + c \qquad \int \frac{f'(x)}{f(x)}\,dx = \ln|f(x)| + c$$

2 *Rational functions*

With the aid of partial fractions all proper rational functions can be reduced to the sum of terms of the form

(i) $\dfrac{a}{bx + c}$ which has an antiderivative of $\dfrac{a}{b}\ln|bx + c|$

(ii) $\dfrac{a}{(bx + c)^2}$ which has an antiderivative of $-\dfrac{a}{b(bx + c)}$

(iii) $\dfrac{ax}{bx^2 + c}$ which has an antiderivative of $\dfrac{a}{2b}\ln|bx^2 + c|$

(iv) $\dfrac{a}{bx^2 + c}$ which has an antiderivative of $\sqrt{\dfrac{b}{c}}\,\dfrac{a}{b}\tan^{-1}\left(\dfrac{x}{\sqrt{\dfrac{c}{b}}}\right)$

(v) $\dfrac{ax + b}{bx^2 + cx + d}$ which is not covered by this course.

Improper fractions can be reduced by algebraic division to the sum of a polynomial and a proper rational function. You should not learn these forms but should be able to derive them in particular cases.

3 *Integration by parts*

Given two functions $f'(x)$ and $g(x)$, the integral of their product can often be simplified by the application of the following rule:

$$\int f'(x)g(x)\,dx = f(x)g(x) - \int f(x)g'(x)\,dx$$

The rule may have to be applied more than once before the integral can be deduced.

4 *Differential equations (variables separable)*

(i) $\dfrac{dy}{dx} = f(x) \Rightarrow \int dy = \int f(x)\,dx \Rightarrow y = \int f(x)\,dx + c$

(ii) $\dfrac{dy}{dx} = f(y) \Rightarrow \int \dfrac{dy}{f(y)} = \int dx \Rightarrow \int \dfrac{dy}{f(y)} = x + c$

(iii) $\dfrac{dy}{dx} = f(x)g(y) \Rightarrow \int \dfrac{dy}{g(y)} = \int f(x)\,dx$

5 Differential equations can be used to model many situations. The topic has applications in modelling growth and decay.

4 Complex Numbers

Historical note

When two whole numbers are added, the result is another whole number. The whole numbers are said to be *closed* under addition.

$$a, b \in W \Rightarrow a + b \in W$$

However, when two whole numbers are subtracted, the result is not always another whole number. Consider $2 - 3 = -1$: not a whole number. The whole numbers are not closed under subtraction.

Diophantus (around AD275) called equations which produced negative numbers *absurd*. By the sixteenth century, negative numbers were perfectly acceptable, and the set of whole numbers was extended, by the inclusion of negative numbers, to the set of integers.

Consider 2 and 3 as integers, $2 - 3 = -1$: an integer.
The integers are closed under subtraction.

$$a, b \in Z \Rightarrow a - b \in Z$$

Consider closure under division and you will see the need for the set of rational numbers. Consider closure when taking the square root and you will see the need for the set of real numbers.

Jerome Cardan

In 1545 Jerome Cardan tried to find the solution to the problem:
What two numbers have a sum of 10 and a product of 40?
His solution involved what he termed as *fictitious* numbers, what we now call *complex numbers.*

In 1637, Descartes used the expressions *real* and *imaginary* in this context and in 1748 Euler used the letter i to stand for the root of the equation $x^2 = -1$.

If we wish to work with $\sqrt{-1}$ we shall need to extend the set of real numbers.

Definitions

- i is a number such that $i^2 = -1$, $i \notin R$.
- C is the set of numbers z, of the form $z = a + ib$ where a and b are real numbers.
- The members of C are called complex numbers.
- a is called the real part of z and we write $a = \mathcal{R}(z)$ or $a = \text{Re}(z)$.
  b is called the imaginary part of z and we write $b = \mathcal{I}(z)$ or $b = \text{Im}(z)$.

- Given $z_1 = a + bi$ and $z_2 = c + di$, addition is defined by

$$z_1 + z_2 = (a + bi) + (c + di) = (a + c) + (b + d)i$$

and multiplication is defined by

$$z_1z_2 = (a + bi)(c + di) = ac + adi + bci + bdi^2$$
$$= (ac - bd) + (ad + bc)i \qquad \text{since } i^2 = -1$$

- If $z_1 = z_2$ then $\mathcal{R}(z_1) = \mathcal{R}(z_2)$ and $\mathcal{I}(z_1) = \mathcal{I}(z_2)$.
This is referred to as *equating real and imaginary parts.*

Proof

Suppose $z_1 = a + bi$ and $z_2 = c + di$ are equal but $a \neq c$ and $b \neq d$.
Then

$$a + bi = c + di$$
$$\Rightarrow \quad a - c = (d - b)i$$
$$\Rightarrow \quad \frac{a - c}{d - b} = i \qquad \text{we can divide since } d - b \neq 0$$
$$\Rightarrow \quad i \in R \qquad \text{(contradiction)}$$
$$\Rightarrow \quad d - b = 0 \quad \Rightarrow \quad d = b$$
$$\text{and } \quad a - c = 0 \quad \Rightarrow \quad a = c$$

Note
We may write $a + ib$ or $a + bi$, whichever we find more convenient.

Addition, subtraction and multiplication

Example 1 Given $z_1 = 3 + 2i$ and $z_2 = 4 + 3i$, find
a $z_1 + z_2$ **b** $z_1 - z_2$ **c** z_1z_2

a $z_1 + z_2 = 3 + 2i + 4 + 3i = 7 + 5i$

b $z_1 - z_2 = 3 + 2i - (4 + 3i) = 3 + 2i - 4 - 3i = -1 - i$

c $z_1z_2 = (3 + 2i)(4 + 3i) = 12 + 9i + 8i + 6i^2 = 12 + 9i + 8i - 6$
$= 6 + 17i$

Example 2 Solve the equation $z^2 - 2z + 5 = 0$.

Using the quadratic formula we get

$$z = \frac{-(-2) \pm \sqrt{(-2)^2 - 4 \times 1 \times 5}}{2 \times 1} = \frac{2 \pm \sqrt{-16}}{2} = 1 \pm \sqrt{-4}$$
$$= 1 \pm \sqrt{4} \times \sqrt{-1}$$
$$= 1 \pm 2i$$

EXERCISE 1

1 Given $z_1 = 2 + i$ and $z_2 = 3 + 4i$ calculate each of the following in the form $a + bi$.

a $z_1 + z_2$ **b** $z_1 z_2$ **c** $3z_1$ **d** $2z_2$

e $4z_1 + 3z_2$ **f** z_1^2 **g** z_1^3 **h** $z_1^3 z_2$

i $-z_2$ **j** $z_1 - z_2$ **k** $z_2 - z_1$ **l** $z_1^2 - 2z_2$

2 Simplify the following, expressing your answers in the form $a + bi$.

a $(3 + 4i) + (1 + i)$ **b** $(6 - 2i) + (4 + 2i)$ **c** $(1 + i)(1 - i)$

d $(1 + 2i)(1 - 3i)$ **e** $(4 - 3i)^2$ **f** $2(3 - i) - 4(1 + 2i)$

g $3(1 + i) - i(1 + 3i)$ **h** $2i(2 + 3i)(1 - 2i)$ **i** $(3 - i)^2(3 + i)$

3 Solve the following quadratic equations giving the roots in the form $z = a \pm bi$.

a $z^2 + 2z + 2 = 0$ **b** $z^2 + 4z + 13 = 0$ **c** $z^2 - 6z + 13 = 0$

d $2z^2 - 4z + 10 = 0$ **e** $3z^2 - 12z + 15 = 0$ **f** $2z^2 + 12z + 36 = 0$

4 Every quadratic equation can be written in the form

$$z^2 - (\textit{sum of the roots})z + (\textit{product of the roots}) = 0$$

Verify this statement for the equations in question 3.

5 Solve Cardan's problem, namely:
Find two numbers which have a sum of 10 and a product of 40.

6 **a** Simplify each of the following.
(i) $(3 + i)(3 - i)$ **(ii)** $(2 + 3i)(2 - 3i)$ **(iii)** $(1 + 2i)(1 - 2i)$

b Comment on your answers in each case. Make a conjecture.

c Simplify $(a + ib)(a - ib)$ to prove your conjecture.

7 **a** $i = i$, $i^2 = -1$, $i^3 = i \times i^2 = -i$, $i^4 = i^2 \times i^2 = 1$
Work out the powers of i up to i^{12}.

b Given that n is an integer, evaluate:
(i) i^{4n-1} **(ii)** i^{4n+1} **(iii)** i^{4n+2} **(iv)** i^{4n} **(v)** i^{4n+3}

8 By equating real and imaginary parts, find a and b in each case.

a $a + bi = (3 + i)^2$ **b** $a + bi = (3 + 2i)^2$ **c** $a + bi = (2 + i)(3 + 4i)$

The complex conjugate

You should have noticed in question 3 of Exercise 1 that the roots of the equations came in pairs of the form $a + bi$ and $a - bi$ where a and b are real numbers. Such pairs are called *complex conjugates.*

You should also have noticed that the product of this pair of complex numbers is in fact a real number, $a^2 + b^2$.

When $z = a + bi$ then its complex conjugate is denoted by $\bar{z} = a - bi$.

$$z\bar{z} = a^2 + b^2$$

This property is very useful when we wish to do a division.

Division and square roots

Example 1 Calculate $(4 + 2i) \div (2 + 3i)$.

$$\frac{(4 + 2i)}{(2 + 3i)} = \frac{(4 + 2i)(2 - 3i)}{(2 + 3i)(2 - 3i)} = \frac{8 - 12i + 4i - 6i^2}{2^2 + 3^2} = \frac{14 - 8i}{13} = \frac{14}{13} - \frac{8}{13}i$$

Note how the complex conjugate is used to make the denominator real.

Example 2 Calculate $\sqrt{5 + 12i}$.

Let $a + bi = \sqrt{5 + 12i}$ where a and b are real. Then

$$(a + bi)^2 = 5 + 12i$$
$$\Rightarrow a^2 - b^2 + 2abi = 5 + 12i$$

Equating real parts, we get

$$a^2 - b^2 = 5 \qquad (1)$$

Equating imaginary parts, we get

$$2ab = 12 \qquad (2)$$

Equation (2) gives

$$a = \frac{6}{b}$$

Substituting into equation (1) gives

$$\left(\frac{6}{b}\right)^2 - b^2 = 5$$

Multiply throughout by b^2.

$$36 - b^4 = 5b^2$$

Rearranging, we get

$$b^4 + 5b^2 - 36 = 0$$

Treating this as a quadratic in b^2, we get

$$b^2 = \frac{-5 \pm \sqrt{25 + 144}}{2} = 4 \text{ or } -9$$

Since $b \in R$,

$$b = \pm\sqrt{4} = 2 \text{ or } -2 \text{ and so } a = 3 \text{ or } -3$$

Thus

$$\sqrt{5 + 12i} = 3 + 2i \text{ or } -3 - 2i$$

EXERCISE 2

1 Calculate the following divisions, expressing your answer in the form $a + ib$ where $a, b \in R$.

a $(8 + 4i) \div (1 + 3i)$ **b** $(8 + i) \div (3 + 2i)$ **c** $(6 + 2i) \div (4 - 2i)$
d $(-1 - 3i) \div (1 - 2i)$ **e** $8 \div (1 + 2i)$ **f** $(6 + i) \div (3 - i)$

2 In each case below, express z^{-1} in the form $a + ib$ where $a, b \in R$.

a $z = i$ **b** $z = 1 - i$ **c** $z = 2 + 2i$
d $z = 3 + i$ **e** $z = 4 - 2i$

3 Simplify:

a $\dfrac{17 - 7i}{5 + i}$ **b** $\dfrac{21 + 9i}{2 + 5i}$ **c** $\dfrac{7 - 3i}{1 + i}$

d $\dfrac{2 - 5i}{1 + i}$ **e** $\dfrac{3 - 2i}{1 + 2i}$ **f** $\dfrac{3}{3 + 4i}$

4 Find a and b in each case so that $(a + ib)^2$ is equal to:

a $5 - 12i$ **b** $15 - 8i$ **c** $-24 - 10i$

5 Calculate:

a $\sqrt{3 - 4i}$ **b** $\sqrt{21 - 20i}$ **c** $\sqrt{-9 + 40i}$

6 **a** If $z = 2 + 3i$ find, in the form $x + iy$:

(i) $\bar{z}$ **(ii)** $\dfrac{1}{\bar{z}}$ **(iii)** $\dfrac{z}{\bar{z}}$ **(iv)** $\dfrac{\bar{z}}{z}$ **(v)** $\dfrac{z}{\bar{z}} + \dfrac{\bar{z}}{z}$ **(vi)** $\dfrac{z}{\bar{z}} - \dfrac{\bar{z}}{z}$

b Repeat **a** when $z = a + bi$.

7 **a** Show that $\frac{1}{2}(z + \bar{z}) = \mathcal{R}(z)$.

b Find a similar expression for $\mathcal{I}(z)$.

8 Given that $z_1 = a + bi$ and $z_2 = x + iy$,

a find expressions for **(i)** $\bar{z}_1$, **(ii)** $\bar{z}_2$ **(iii)** $\overline{z_1 + z_2}$;

b state the simple relationship between $\bar{z}_1$, $\bar{z}_2$ and $\overline{z_1 + z_2}$;

c identify similar conclusions for **(i)** $\overline{z_1 - z_2}$, **(ii)** $\overline{z_1 \times z_2}$, **(iii)** $\overline{z_1 \div z_2}$.

A geometric interpretation: Argand diagrams

At the end of the eighteenth century, Caspar Wessel from Norway and Jean Robert Argand from Switzerland independently came up with a geometric interpretation of the complex number $z = x + iy$.

The complex number $z = x + iy$ is represented on the plane by the point P(x, y). The plane is referred to as the *complex plane*, and diagrams of this sort are often called *Argand diagrams*.

Any point on the x-axis represents a purely real number.
Any point on the y-axis represents a purely imaginary number.

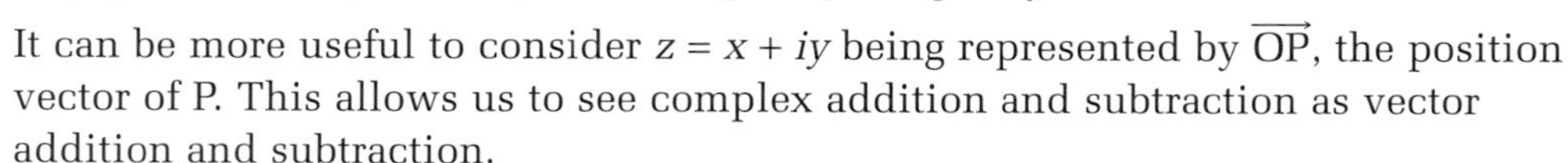
It can be more useful to consider $z = x + iy$ being represented by $\overrightarrow{OP}$, the position vector of P. This allows us to see complex addition and subtraction as vector addition and subtraction.

$\overrightarrow{OP}$ is considered a vector which has been rotated off the x-axis. The length of $\overrightarrow{OP}$, r, is called the *modulus of* z and is denoted by $|z|$.

The size of the rotation is called the *amplitude* or *argument* of z. It is often denoted by Arg z. This angle of rotation could be $\theta \pm 2n\pi$ where n is any integer. We refer to the value of Arg z which lies in the range $-\pi < \theta \leq \pi$ as the *principal argument*. It is denoted by arg z. [Note the lower case a.]

From the diagram above, by Pythagoras' theorem:

$$r = \sqrt{x^2 + y^2}$$

By simple trigonometry:

$$\theta = \tan^{-1}\left(\frac{y}{x}\right), \quad -\pi < \theta \leq \pi$$

By simple trigonometry:

$$x = r\cos\theta$$
$$y = r\sin\theta$$

Thus $z = x + iy$ can be re-written as

$$z = r\cos\theta + ir\sin\theta$$

i.e.

$$z = r(\cos\theta + i\sin\theta)$$

This is referred to as the *polar form* of z.

Example 1 Find the modulus and argument of the complex number $z = 3 + 4i$.

$$|z| = \sqrt{3^2 + 4^2} = 5$$

$$\text{Arg } z = \tan^{-1}\left(\frac{4}{3}\right) + n\pi = 0.927 + n\pi \text{ radians}$$

Check: (3, 4) is in the first quadrant so $n = 0$.

$$\arg z = 0.927 \text{ radians (3 s.f.)}$$

Example 2 Find the modulus and argument of the complex number $z = -3 - 4i$.

$$|z| = \sqrt{(-3)^2 + (-4)^2} = 5$$

$$\text{Arg } z = \tan^{-1}\left(\frac{-4}{-3}\right) + n\pi = 0.927 + n\pi \text{ radians.}$$

Check: (–3, –4) is in the fourth quadrant so $n = -1$.

$$\arg z = 0.927 - \pi = -2.21 \text{ (3 s.f.)}$$

Example 3 Express $z = 2 + 2i$ in the form $r(\cos\theta° + i\sin\theta°)$.

$$r = |z| = \sqrt{2^2 + 2^2} = 2\sqrt{2}$$

$$\theta = \arg z = \tan^{-1}\left(\frac{2}{2}\right) = 45° \text{ (check quadrant)}$$

Hence $z = 2\sqrt{2}(\cos 45° + i\sin 45°)$.

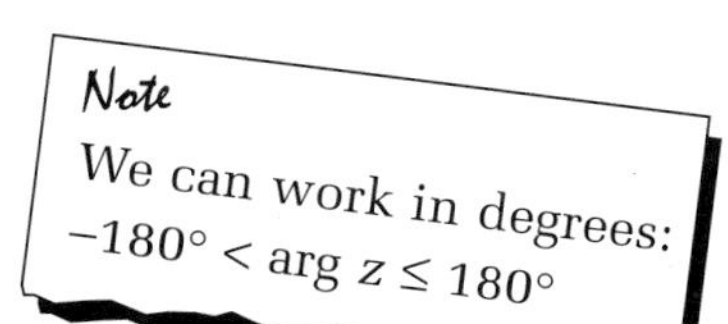

EXERCISE 3

1 $z = 1 \Rightarrow |z| = 1$ and $\arg z = 0$. Use the diagram to help you make similar statements about

a $z = i$

b $z = -1$

c $z = -i$

2 a Draw Argand diagrams to illustrate:

(i) $3 + 4i$ and $3 - 4i$ **(ii)** $2 + 3i$ and $2 - 3i$ **(iii)** $5 + i$ and $5 - i$

b Comment on the Argand diagrams of z and $\overline{z}$.

3 For each of the following complex numbers,

(i) plot the number on an Argand diagram,

(ii) find the modulus and argument to three significant figures where appropriate.

a $1 + i$ **b** $2 + 3i$ **c** $3 + 2i$

d 6 **e** $3i$ **f** $-4 - 3i$ (refer to your sketch)

g $-1 + 2i$ **h** $2 - 3i$ **i** $4 - i$

4 For each of the following expressions,

(i) simplify by writing in the form $x + iy$,

(ii) find the modulus and argument.

a $\dfrac{3 + 2i}{1 + 5i}$ **b** $\dfrac{1}{1 + 3i}$ **c** $(2 + 4i)(1 - i)$

5 a Find the modulus and argument of each expression.

(i) $10 + 7i$ **(ii)** $(10 + 7i)^2$ **(iii)** $(10 + 7i)^3$

b Comment on any connection you see.

6 Find the complex number with:

a $|z| = 2, \arg z = \dfrac{\pi}{6}$ **b** $|z| = 3, \arg z = \dfrac{\pi}{4}$ **c** $|z| = 4, \arg z = \dfrac{\pi}{2}$

d $|z| = 3, \arg z = \dfrac{\pi}{3}$ **e** $|z| = 2, \arg z = -\dfrac{\pi}{4}$ **f** $|z| = 1, \arg z = -\dfrac{\pi}{6}$

7 Express each of these complex numbers in polar form (give the argument in degrees).

a $1 + i\sqrt{3}$ **b** $\sqrt{2} + i\sqrt{2}$ **c** $-2\sqrt{3} + 2i$

d -1 **e** $3i$ **f** $-4 - i4\sqrt{3}$

g $-2\sqrt{x} - 2i$ **h** $-\sqrt{2} - i\sqrt{2}$ **i** $-1 - i\sqrt{3}$

8 Given that $z_1 = 2 + 3i$ and $z_2 = 3 + i$, the diagram illustrates the sum $z_1 + z_2$, i.e.

$$(2 + 3i) + (3 + i) = 5 + 4i$$

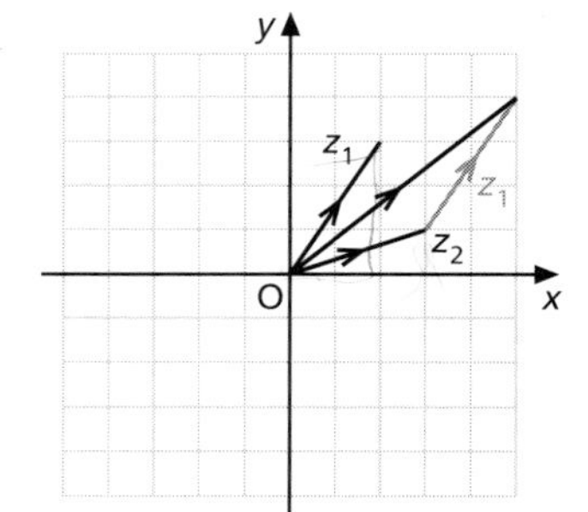

On similar diagrams, illustrate:

a $(2 + 3i) + (3 + 2i)$ **b** $(3 + 3i) + (2 + 2i)$

c $(2 - 3i) + (3 - 2i)$ **d** $(-2 + 3i) + (-3 - 2i)$

e $(2 + 3i) - (3 + 2i)$ [Hint: $= (2 + 3i) + (-3 - 2i)$]

f $(1 + 3i) - (2 + 4i)$ **g** $(-3 + i) - (-1 - 2i)$

h $(-1 - 2i) - (4 - 3i)$ **i** $(-2 - 2i) - (-3 - 3i)$

Sets of points (loci) on the complex plane

Sometimes we have to find the locus of a point which moves in the complex plane with restrictions placed on its modulus and argument.

Example 1 Given that $z = x + iy$, draw the locus of the point which moves on the complex plane so that **a** $|z| = 4$, **b** $|z| \leq 4$.

$$|z| = 4 \Rightarrow \sqrt{x^2 + y^2} = 4 \Rightarrow x^2 + y^2 = 16$$

This is a circle, centre the origin, radius 4.

a

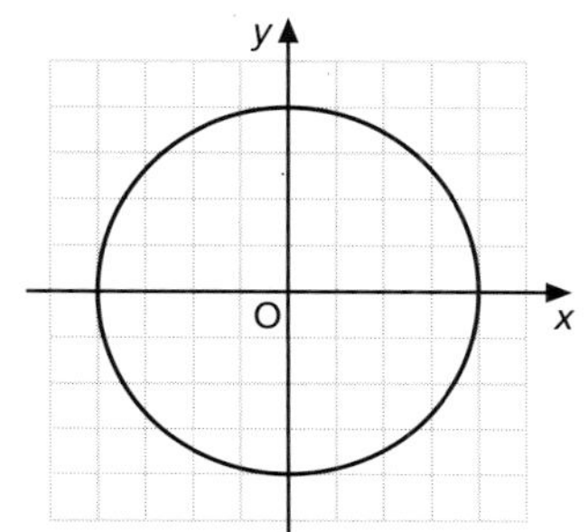

$|z| = 4$

b

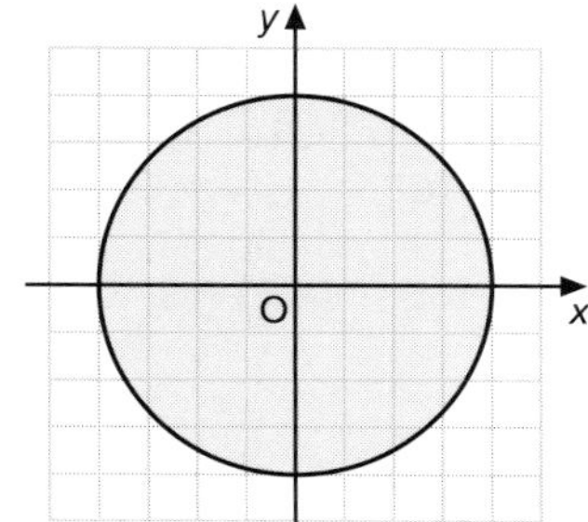

$|z| \leq 4$

Example 2 If $z = x + iy$,

a find the equation of the locus $|z - 2| = 3$,

b draw the locus on an Argand diagram.

a
$$|z - 2| = 3 \Rightarrow |x - 2 + iy| = 3 \Rightarrow \sqrt{(x-2)^2 + y^2} = 3$$
$$\Rightarrow (x-2)^2 + y^2 = 9$$

This is a circle, centre (2, 0) and radius 3.

b

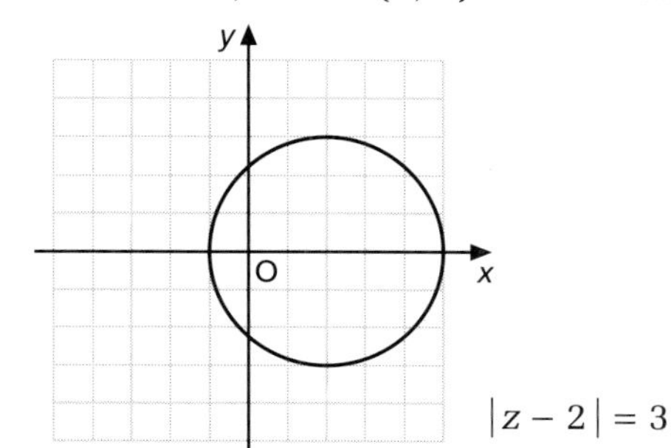

$|z - 2| = 3$

Example 3 If $z = x + iy$, find the equation of the locus $\arg z = \frac{\pi}{3}$.

$$\arg z = \frac{\pi}{3} \Rightarrow \tan^{-1}\left(\frac{y}{x}\right) = \frac{\pi}{3}$$
$$\Rightarrow \frac{y}{x} = \tan\frac{\pi}{3} \Rightarrow \frac{y}{x} = \sqrt{3}$$
$$\Rightarrow y = \sqrt{3}x$$

This is a straight line passing through the origin, gradient $\sqrt{3}$.

EXERCISE 4

1 Given that $z = x + iy$, for each of the following, **(i)** find the equation of the locus, **(ii)** draw the locus on an Argand diagram.

a $|z| = 5$ **b** $|z-3| = 2$ **c** $|z+1| = 4$ **d** $|z+i| = 3$

e $|z-2i| = 3$ **f** $|z+1+2i| = 3$ **g** $|2z+3i| = 5$ **h** $|3z-i| = 5$

i $|3z+3-2i| = 4$ **j** $\arg z = \frac{\pi}{6}$ **k** $\arg z = \frac{\pi}{4}$ **l** $\arg z = \frac{2\pi}{3}$

m $\arg z = 1$ **n** $2 \arg z = \frac{\pi}{4}$ **o** $\arg z = -\frac{\pi}{3}$

2 Explore the loci of the form:

a $|z-a| = b$ **b** $|z-ai| = b$ **c** $|z-ai-b| = c$

3 Given that $z = x + iy$, find the equation of the locus in each case.

a $|z-1| = |z-i|$ **b** $|z-2| = |z-i|$ **c** $|z-3| = |z-2i|$

d $|z-a| = |z-bi|$

4 Sketch the following loci, given that $z = x + iy$.

a $|z| \leq 3$ **b** $|z-3| \leq 2$ **c** $|z+2| \geq 5$

A diversion

Using a spreadsheet or the parametric graphing facility on a calculator you can illustrate more complicated loci where the modulus is expressed as a function of the argument.

	A	B	C	D
1	argument	modulus	x	y
2	-3.14	0.00	0.00	0.00
3	-2.83	0.05	-0.05	-0.02
4	-2.51	0.19	-0.15	-0.11
5	-2.20	0.41	-0.24	-0.33
6	-1.88	0.69	-0.21	-0.66
7	-1.57	1.00	0.00	-1.00
8	-1.26	1.31	0.40	-1.24
9	-0.94	1.59	0.93	-1.28
10	-0.63	1.81	1.46	-1.06
11	-0.31	1.95	1.86	-0.60
12	0.00	2.00	2.00	0.00
13	0.31	1.95	1.86	0.60
14	0.63	1.81	1.46	1.06
15	0.94	1.59	0.93	1.28
16	1.26	1.31	0.40	1.24
17	1.57	1.00	0.00	1.00
18	1.88	0.69	-0.21	0.66
19	2.20	0.41	-0.24	0.33
20	2.51	0.19	-0.15	0.11
21	2.83	0.05	-0.05	0.02
22	3.14	0.00	0.00	0.00

- A2 contains =–PI().
- A3 contains =A2+0.1*PI().
- A3 is *filled down* to A22. This gives us values of the argument, θ, in the range $-\pi < \theta \leq \pi$.
- Where the argument, r, is a function of θ, i.e. $r = f(\theta)$, then $= f(\theta)$ is entered in B2.
- B2 is *filled down* to B22.
- C2 contains =A2*cos(A2), which gives $x = r \cos \theta$.
- D2 contains =A2*sin(A2), which gives $y = r \sin \theta$.
- C2 and D2 are *filled down* to C22 and D22.
- An X-Y chart is then drawn using columns C and D.

Better resolution can be achieved by entering a smaller step size in A3, for example

- A3 contains =A2+0.05*PI().
- A3 is *filled down* to A42.

This example is the locus of the points where $|z| = 1 + \cos(\arg z)$
- B2 contains =1+cos(A2).

The locus is heart-shaped and is called a *cardioid*.

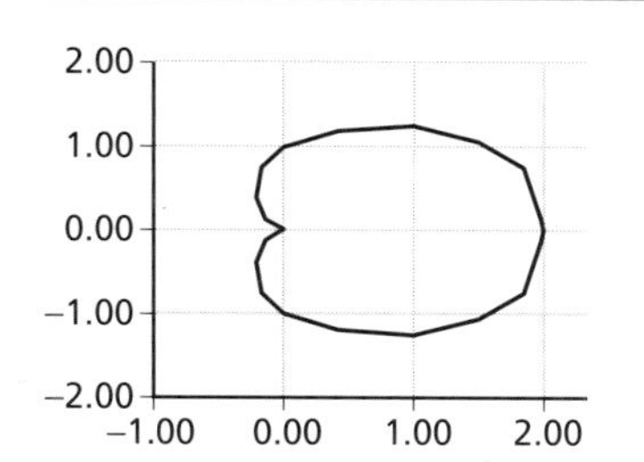

If you have access to a spreadsheet, try entering the following in B2.

1 =cos(2*A2)
  Explore cos(a*A2) for different $a \in N$: *rhodonae (rose curves).*
2 =sin(2*A2)
  Explore sin(a*A2) for different $a \in N$.
3 =1+2*cos(A2) (*the limaçon*)
4 =1+2*cos(2*A2)
5 =A2
  Try this in the range $[0, 2\pi]$ also.
6 =a, where a is a constant (*circles*)
7 =1/(1+0.5*cos(A2)) (*the ellipse*)

On a graphics calculator
- Set your calculator to *parametric plotting*

For spreadsheet exercise 1
- Y1T = cos(2T)cosT
- Y2T = cos(2T)sinT
- Graph.

In general, to draw $|z| = f(\arg z)$,
- Y1T = $f(\arg z)$cosT
- Y2T = $f(\arg z)$sinT

Polar form and multiplication

Recall that $z = r(\cos\theta + i\sin\theta)$ is the *polar form* of z.
Consider z_1z_2 where $z_1 = a(\cos A + i\sin A)$ and $z_2 = b(\cos B + i\sin B)$.

$$\begin{aligned} z_1z_2 &= a(\cos A + i\sin A)b(\cos B + i\sin B) \\ &= ab(\cos A\cos B + \cos A.i\sin B + i\sin A\cos B + i^2\sin A\sin B) \\ &= ab([\cos A\cos B - \sin A\sin B] + i[\cos A\sin B + \sin A\cos B]) \\ z_1z_2 &= ab(\cos(A+B) + i\sin(A+B)) \end{aligned}$$

We see that

$$|z_1z_2| = |z_1| \times |z_2|$$
$$\text{Arg}(z_1z_2) = \text{Arg}\, z_1 + \text{Arg}\, z_2$$

But note that the *principal* argument, $\arg(z_1z_2)$, lies in the range $(-\pi, \pi]$, and adjustments have to be made by adding or subtracting 2π as appropriate if $\text{Arg}(z_1z_2)$ goes outside this range during the calculation.

Notice that multiplying $z = r(\cos\theta + i\sin\theta)$ by $\cos A + i\sin A$ gives $r(\cos(\theta + A) + i\sin(\theta + A))$. This is equivalent to rotating the vector representing z through an angle of A radians anticlockwise. Since $i = \cos\frac{\pi}{2} + i\sin\frac{\pi}{2}$ then multiplying by i is equivalent to rotating z through an angle of $\frac{\pi}{2}$ radians anticlockwise.

If $z = r(\cos\theta + i\sin\theta)$ then

$$\frac{1}{z} = \frac{1}{r(\cos\theta + i\sin\theta)} = \frac{(\cos\theta - i\sin\theta)}{r(\cos\theta + i\sin\theta)(\cos\theta - i\sin\theta)}$$

$$= \frac{1}{r}(\cos\theta - i\sin\theta) = \frac{1}{r}(\cos(-\theta) + i\sin(-\theta))$$

So

$$\left|\frac{1}{z}\right| = \frac{1}{|z|} \text{ and } \arg\left(\frac{1}{z}\right) = -\arg z$$

$$\frac{z_1}{z_2} = z_1 \times \frac{1}{z_2} = \frac{a}{b}[\cos(A - B) + i\sin(A - B)]$$

We see that

$$\left|\frac{z_1}{z_2}\right| = |z_1| \div |z_2|$$

$$\text{Arg}\left(\frac{z_1}{z_2}\right) = \text{Arg } z_1 - \text{Arg } z_2$$

But note that the *principal* argument lies in the range $(-\pi, \pi]$, and adjustments have to be made by adding or subtracting 2π as appropriate if $\text{Arg}\left(\frac{z_1}{z_2}\right)$ goes outside this range during the calculation.

EXERCISE 5

1 Simplify the following.

a $3\left(\cos\frac{\pi}{3} + i\sin\frac{\pi}{3}\right) \times 4\left(\cos\frac{\pi}{2} + i\sin\frac{\pi}{2}\right)$

b $2\left(\cos\frac{\pi}{4} + i\sin\frac{\pi}{4}\right) \times 5\left(\cos\frac{\pi}{6} + i\sin\frac{\pi}{6}\right)$

c $2\left(\cos\frac{\pi}{3} + i\sin\frac{\pi}{3}\right) \times 4\left(\cos\frac{\pi}{3} - i\sin\frac{\pi}{3}\right)$

[Hint: remember $\left(\cos\frac{\pi}{3} - i\sin\frac{\pi}{3}\right) = \left(\cos\left(-\frac{\pi}{3}\right) + i\sin\left(-\frac{\pi}{3}\right)\right)$.]

d $\left(\cos\frac{\pi}{2} - i\sin\frac{\pi}{2}\right) \times 2\left(\cos\frac{\pi}{3} - i\sin\frac{\pi}{3}\right)$

e $5\left(\cos\frac{\pi}{5} - i\sin\frac{\pi}{5}\right) \times 2\left(\cos\frac{\pi}{6} + i\sin\frac{\pi}{6}\right)$

f $4\left(\cos\frac{\pi}{2} + i\sin\frac{\pi}{2}\right) \div 2\left(\cos\frac{\pi}{3} + i\sin\frac{\pi}{3}\right)$

g $5\left(\cos\frac{\pi}{4} + i\sin\frac{\pi}{4}\right) \div 2\left(\cos\frac{\pi}{8} + i\sin\frac{\pi}{8}\right)$

h $9\left(\cos\frac{\pi}{2} + i\sin\frac{\pi}{2}\right) \div 3\left(\cos\frac{\pi}{6} - i\sin\frac{\pi}{6}\right)$

i $8\left(\cos\frac{\pi}{7} - i\sin\frac{\pi}{7}\right) \div 2\left(\cos\frac{2\pi}{7} - i\sin\frac{2\pi}{7}\right)$

2 Convert each complex number to polar form then state the product, z_1z_2, and quotient, $\frac{z_1}{z_2}$, of each pair. (Work to three significant figures.)

a $z_1 = 3 + 4i, \quad z_2 = 1 + i$
b $z_1 = 2 + 3i, \quad z_2 = 3 - i$
c $z_1 = 1 - i, \quad z_2 = -1 - i$
d $z_1 = -4 + i, \quad z_2 = -2 + 2i$

3 Given that $z = r\left(\cos\frac{\pi}{3} + i\sin\frac{\pi}{3}\right)$ calculate:

a $z^2 \left[= r^2\left(\cos\frac{\pi}{3} + i\sin\frac{\pi}{3}\right)^2 = r^2\left(\cos\frac{\pi}{3} + i\sin\frac{\pi}{3}\right)\left(\cos\frac{\pi}{3} + i\sin\frac{\pi}{3}\right)\right]$

b $z^3 \quad [= z^2z]$

c z^4 [Remember to bring the argument back into the range $(-\pi, \pi]$.]

d z^5 **e** z^6 **f** z^7

4 Repeat question 3 using

(i) $z = r\left(\cos\frac{\pi}{2} + i\sin\frac{\pi}{2}\right)$
(ii) $z = r\left(\cos\frac{2\pi}{3} + i\sin\frac{2\pi}{3}\right)$

(iii) $z = r\left(\cos\frac{3\pi}{4} + i\sin\frac{3\pi}{4}\right)$
(iv) $z = r(\cos\theta + i\sin\theta)$

5 a Use the binomial theorem to expand $(3 + 4i)^4$.

b (i) Express $z = 3 + 4i$ in polar form (to 3 s.f.).

(ii) Use the notion developed in questions 3 and 4 to find $z^4 = (3 + 4i)^4$ in polar form.

(iii) Use your calculator to help you express z^4 in the form $a + bi$.

c Which method, **a** or **b**, would be preferable when calculating $(3 + 4i)^{20}$?

De Moivre's theorem

Abraham De Moivre

Questions 3 and 4 above suggest that if $z = r(\cos\theta + i\sin\theta)$ then $z^n = r^n(\cos n\theta + i\sin n\theta)$.

The development of this idea is attributed to Abraham De Moivre (1667–1754). The result is called De Moivre's theorem, although it was Euler who proved it true for all $n \in R$.

Proof for $n \in N$

Suppose that there exists a natural number $n = k$ for which the theorem is true. Then, given $z = r(\cos\theta + i\sin\theta)$, we have $z^k = r^k(\cos k\theta + i\sin k\theta)$

$$\begin{aligned} z^{k+1} &= r^k(\cos k\theta + i\sin k\theta) \times r(\cos\theta + i\sin\theta) \\ &= r^{k+1}(\cos(k\theta + \theta) + i\sin(k\theta + \theta)) \\ &= r^{k+1}(\cos(k+1)\theta + i\sin(k+1)\theta) \end{aligned}$$

Thus, if the theorem is true for $n = k$, then it is true for $n = k + 1$.
We know, however, that it *is* true for $n = 1$.
Thus, by induction, it is true for all $n \geq 1, n \in N$.

Proof for all $n \in R$ is beyond the scope of this course but it can be illustrated and verified by suitable examples.

For example, given $z = r(\cos\theta + i\sin\theta)$ and $z_1 = \sqrt{r\left(\cos\frac{\theta}{2} + i\sin\frac{\theta}{2}\right)}$ then

$$z_1^2 = (\sqrt{r})^2\left(\cos\frac{\theta}{2} + i\sin\frac{\theta}{2}\right)^2 = r\left(\cos\frac{2\theta}{2} + i\sin\frac{2\theta}{2}\right) \quad \text{by De Moivre's theorem}$$
$$= r(\cos\theta + i\sin\theta) = z$$
$$z_1 = \sqrt{z}$$

It would appear that De Moivre's theorem holds for $n = \frac{1}{2}$ i.e.

$$[r(\cos\theta + i\sin\theta)]^{\frac{1}{2}} = r^{\frac{1}{2}}\left(\cos\frac{\theta}{2} + i\sin\frac{\theta}{2}\right)$$

Can you show that it holds when $n = -1$?

Example 1 Given $z = 1 + i\sqrt{3}$, find **a** z^2, **b** z^5, **c** z^7, **d** z^{10}.

$$|z| = \sqrt{1^2 + (\sqrt{3})^2} = 2, \quad \arg z = \tan^{-1}\left(\frac{\sqrt{3}}{1}\right) = \frac{\pi}{3}, \quad z = 2\left(\cos\frac{\pi}{3} + i\sin\frac{\pi}{3}\right)$$

a $z^2 = 2^2\left(\cos\frac{\pi}{3} + i\sin\frac{\pi}{3}\right)^2 = 4\left(\cos\frac{2\pi}{3} + i\sin\frac{2\pi}{3}\right)$

$$= 4\left(-\frac{1}{2} + i\frac{\sqrt{3}}{2}\right) = -2 + i2\sqrt{3}$$

b $z^5 = 2^5\left(\cos\frac{5\pi}{3} + i\sin\frac{5\pi}{3}\right)$

Argument must be brought into range $(-\pi, \pi]$ by subtracting 2π.

$$= 32\left(\cos\left(-\frac{\pi}{3}\right) + i\sin\left(-\frac{\pi}{3}\right)\right) = 32\left(\cos\frac{\pi}{3} - i\sin\frac{\pi}{3}\right)$$
$$= 32\left(\frac{1}{2} - i\frac{\sqrt{3}}{2}\right) = 16 - i\,16\sqrt{3}$$

c $z^7 = 2^7\left(\cos\frac{7\pi}{3} + i\sin\frac{7\pi}{3}\right)$

Argument must be brought into range $(-\pi, \pi]$ by subtracting 2π.

$$= 128\left(\cos\left(\frac{\pi}{3}\right) + i\sin\left(\frac{\pi}{3}\right)\right)$$
$$= 128\left(\frac{1}{2} + i\frac{\sqrt{3}}{2}\right) = 64 - i\,64\sqrt{3}$$

d $z^{10} = 2^{10}\left(\cos\frac{10\pi}{3} + i\sin\frac{10\pi}{3}\right)$

Argument must be brought into range $(-\pi, \pi]$ by subtracting 2π twice.

$$= 1024\left(\cos\left(-\frac{2\pi}{3}\right) + i\sin\left(-\frac{2\pi}{3}\right)\right) = 1024\left(\cos\frac{2\pi}{3} - i\sin\frac{2\pi}{3}\right)$$
$$= 1024\left(-\frac{1}{2} - i\frac{\sqrt{3}}{2}\right) = -512 - i\,512\sqrt{3}$$

Example 2 Given $z = 2 + i$, calculate z^4.
[Work to three decimal places and round the final answer to the nearest integer.]

$$|z| = \sqrt{2^2 + 1^2} = \sqrt{5}, \quad \arg z = \tan^{-1}\left(\frac{1}{2}\right) = 0.464 \text{ (3 d.p.)},$$
$$z = \sqrt{5}(\cos 0.464 + i \sin 0.464)$$
$$z^4 = (\sqrt{5}(\cos 0.464 + i \sin 0.464))^4 = 25(\cos 1.856 + i \sin 1.856) \text{ to 3 d.p.}$$
$$= 25(-0.281 + i \times 0.960)$$
$$= -7 + 24i \quad \text{(to nearest integer)}$$

EXERCISE 6

1 For each of the following complex numbers z,
- express it in polar form,
- find each of the required powers in polar form, bringing the argument into the range $(-\pi, \pi]$,
- finally express your answers in the form $a + ib$.

a Given $z = 2\sqrt{3} + 2i$ find (i) z^2 (ii) z^5 (iii) z^{10}
b Given $z = \sqrt{3} - i$ find (i) z^3 (ii) z^4 (iii) z^8
c Given $z = 1 + i$ find (i) z^3 (ii) z^6 (iii) z^{12}

2 Simplify, giving your answers correct to three significant figures:

a $\left[3\left(\cos\frac{\pi}{5} + i \sin\frac{\pi}{5}\right)\right]^3$ **b** $\left[2\left(\cos\frac{\pi}{6} + i \sin\frac{\pi}{6}\right)\right]^8$

c $\left(\cos\frac{\pi}{4} + i \sin\frac{\pi}{4}\right)^2\left(\cos\frac{3\pi}{4} + i \sin\frac{3\pi}{4}\right)^2$ **d** $\left(\cos\frac{2\pi}{7} + i \sin\frac{2\pi}{7}\right)^3\left(\cos\frac{3\pi}{7} + i \sin\frac{3\pi}{7}\right)^4$

3 In each of the following, work to three decimal places where necessary then round the components of your final answer to the nearest whole number.

a Given $z = 2 + 3i$, calculate z^3. **b** Given $z = -1 + 4i$, calculate z^5.
c Given $z = -2 - 3i$, calculate z^4. **d** Given $z = 2 - 2i$, calculate z^7.

4 The argument can be given in degrees. The same laws apply and the range of the argument is $(-180°, 180°]$. Simplify the following:

a $(\cos 10° + i \sin 10°)(\cos 30° + i \sin 30°)$
b $(\cos 50° + i \sin 50°)(\cos 145° + i \sin 145°)$
c $(\cos 25° + i \sin 25°)(\cos 20° - i \sin 20°)$
d $(\cos 150° - i \sin 150°)(\cos 40° - i \sin 40°)$
e $(\cos 30° + i \sin 30°) \div (\cos 10° + i \sin 10°)$
f $(\cos 4° + i \sin 4°) \div (\cos 10° + i \sin 10°)$
g $(\cos 20° + i \sin 20°)^3(\cos 30° + i \sin 30°)^2$
h $(\cos 125° - i \sin 125°)^4(\cos 15° - i \sin 15°)^3$

i $\dfrac{(\cos 40° + i \sin 40°)^3}{(\cos 10° + i \sin 10°)^2}$ **j** $\dfrac{(\cos 6° + i \sin 6°)^5}{(\cos 3° + i \sin 3°)^2}$

k $\dfrac{(\cos 25° + i \sin 25°)^4}{(\cos 7° + i \sin 7°)(\cos 3° + i \sin 3°)}$ **l** $\dfrac{(\cos 32° + i \sin 32°)^4}{(\cos 16° + i \sin 16°)^3(\cos 4° - i \sin 4°)^2}$

5 **a** Expand $(\cos\theta + i\sin\theta)^2$ **(i)** using the binomial theorem,
(ii) using De Moivre's theorem.

b **(i)** By equating the *real* parts, express $\cos 2\theta$ in terms of $\sin\theta$ and $\cos\theta$.
(ii) By equating the *imaginary* parts, express $\sin 2\theta$ in terms of $\sin\theta$ and $\cos\theta$.

6 **a** Expand $(\cos\theta + i\sin\theta)^3$ **(i)** using the binomial theorem,
(ii) using De Moivre's theorem.

b **(i)** By equating the *real* parts, express $\cos 3\theta$ in terms of $\sin\theta$ and $\cos\theta$.
(ii) Use the identity $\sin^2\theta + \cos^2\theta = 1$ to help you express $\cos 3\theta$ in terms of $\cos\theta$ only.

c Express $\sin 3\theta$ in terms of $\sin\theta$.

d Hence express $\sin^3\theta$ in terms of $\sin\theta$ and $\sin 3\theta$.

7 **a** By considering the expansion of $(\cos\theta + i\sin\theta)^4$, express:
(i) $\cos 4\theta$ in terms of $\cos\theta$
(ii) $\sin 4\theta$ in terms of $\sin\theta$ and $\cos\theta$
(iii) $\cos^4\theta$ in terms of $\cos\theta$ and $\cos 4\theta$

b Express:
(i) $\cos 5\theta$ in terms of $\cos\theta$
(ii) $\sin 5\theta$ in terms of $\sin\theta$
(iii) $\cos^5\theta$ in terms of $\cos\theta$ and $\cos 5\theta$

8 $z_1 = \cos\frac{11\pi}{6} - i\sin\frac{11\pi}{6}$, $z_2 = \cos\frac{5\pi}{6} - i\sin\frac{5\pi}{6}$, $z_3 = \cos\frac{\pi}{6} + i\sin\frac{\pi}{6}$, $z_4 = \cos\frac{7\pi}{6} + i\sin\frac{7\pi}{6}$.

a Find expressions for **(i)** z_1^2, **(ii)** z_2^2, **(iii)** z_3^2 **(iv)** z_4^2.

b **(i)** Reduce each argument so that it lies in the range $(-\pi, \pi]$.
(ii) How many distinct answers are obtained?

c **(i)** Reduce the arguments of z_1, z_2, z_3 and z_4.
(ii) If asked for $z \in C$, such that $z^2 = \left(\cos\frac{\pi}{3} + i\sin\frac{\pi}{3}\right)$, what would be a complete answer?
(iii) Illustrate the set of solutions on an Argand diagram.

9 $z_1 = \cos\frac{13\pi}{12} - i\sin\frac{13\pi}{12}$, $z_2 = \cos\frac{5\pi}{12} - i\sin\frac{5\pi}{12}$, $z_3 = \cos\frac{\pi}{4} + i\sin\frac{\pi}{4}$, $z_4 = \cos\frac{11\pi}{12} + i\sin\frac{11\pi}{12}$, $z_5 = \cos\frac{19\pi}{12} + i\sin\frac{19\pi}{12}$.

a Verify in each case that $z^3 = \cos\frac{3\pi}{4} + i\sin\frac{3\pi}{4}$.

b **(i)** Reduce the arguments of z_1, z_2, z_3, z_4 and z_5 to lie in the range $(-\pi, \pi]$.
(ii) If asked for $z \in C$, such that $z^3 = \left(\cos\frac{3\pi}{4} + i\sin\frac{3\pi}{4}\right)$, what would be a complete answer?
(iii) Illustrate the set of solutions on an Argand diagram.

Roots of a complex number

Question 8 of Exercise 6 explored solutions of the equation $z^2 = \left(\cos\frac{\pi}{3} + i\sin\frac{\pi}{3}\right)$.

Two solutions were found, namely

$$z_2 = \cos\frac{5\pi}{6} - i\sin\frac{5\pi}{6} \quad \text{and} \quad z_3 = \cos\frac{\pi}{6} + i\sin\frac{\pi}{6}$$

On an Argand diagram we have

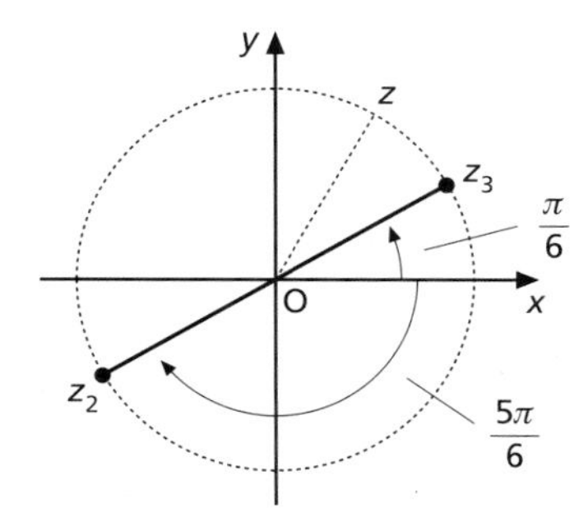

Note

Starting with the position vector of 1 on the x-axis,

- a rotation of $\frac{\pi}{6}$ takes us to z_3 and a rotation of $2 \times \frac{\pi}{6}$ takes us to z,
- a rotation of $-\frac{5\pi}{6}$ takes us to z_2 and a rotation of $2 \times \left(-\frac{5\pi}{6}\right)$ takes us to z,
- the position vectors of the **two** solutions are $\frac{2\pi}{2}$ radians apart.

Check that each solution is of the form

$$z = \left(\cos\left(\frac{\pi}{3} + 2k\pi\right) + i\sin\left(\frac{\pi}{3} + 2k\pi\right)\right)^{\frac{1}{2}} \quad \text{where } k = 0, 1$$

By De Moivre's theorem

$$z = \left(\cos\frac{1}{2}\left(\frac{\pi}{3} + 2k\pi\right) + i\sin\frac{1}{2}\left(\frac{\pi}{3} + 2k\pi\right)\right)$$

$$z = \left(\cos\left(\frac{\pi}{6} + k\pi\right) + i\sin\left(\frac{\pi}{6} + k\pi\right)\right)$$

[Remember to bring the argument into the range $(-\pi, \pi]$.]

Question 9 of Exercise 6 considered solutions of the equation

$$z^3 = \left(\cos\frac{3\pi}{4} + i\sin\frac{3\pi}{4}\right)$$

Three solutions were found, namely

$$z_2 = \cos\frac{5\pi}{12} - i\sin\frac{5\pi}{12}, \quad z_3 = \cos\frac{\pi}{4} + i\sin\frac{\pi}{4}, \quad \text{and} \quad z_4 = \cos\frac{11\pi}{12} + i\sin\frac{11\pi}{12}$$

On an Argand diagram we have

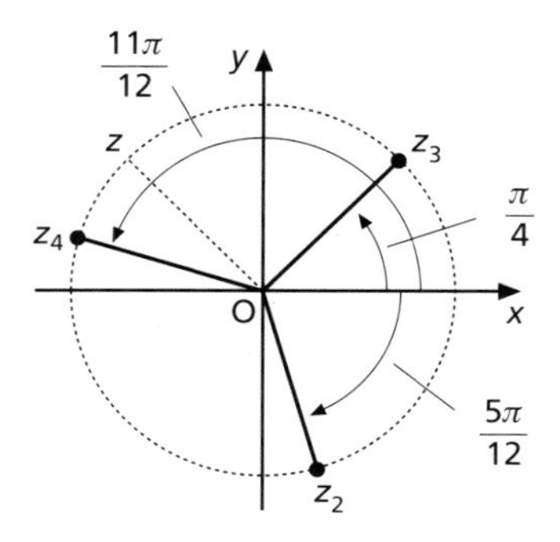

Note

Starting with the position vector of 1 on the x-axis,

- a rotation of $\frac{\pi}{4}$ takes us to z_3 and a rotation of $3 \times \frac{\pi}{4}$ takes us to z,
- a rotation of $-\frac{5\pi}{12}$ takes us to z_2 and a rotation of $3 \times \left(-\frac{5\pi}{12}\right)$ takes us to z,
- a rotation of $\frac{11\pi}{12}$ takes us to z_4 and a rotation of $3 \times \frac{11\pi}{12}$ takes us to z,
- the position vectors of the **three** solutions are $\frac{2\pi}{3}$ radians apart.

Check that each solution is of the form

$$z = \left(\cos\left(\frac{3\pi}{4} + 2k\pi\right) + i\sin\left(\frac{3\pi}{4} + 2k\pi\right)\right)^{\frac{1}{3}} \quad \text{where } k = 0, 1, 2$$

By De Moivre's theorem

$$z = \left(\cos\frac{1}{3}\left(\frac{3\pi}{4} + 2k\pi\right) + i\sin\frac{1}{3}\left(\frac{3\pi}{4} + 2k\pi\right)\right)$$

$$z = \left(\cos\left(\frac{\pi}{4} + \frac{2k\pi}{4}\right) + i\sin\left(\frac{\pi}{4} + \frac{2k\pi}{4}\right)\right)$$

[Remember to bring the argument into the range $(-\pi, \pi]$.]

By De Moivre's theorem, when finding the nth root of a complex number we are effectively dividing the argument by n. We should therefore study arguments in the range $(-n\pi, n\pi]$ so that we have all the solutions in the range $(-\pi, \pi]$ after division by n.

If $z = r(\cos\theta + i\sin\theta)$ then the n solutions of the equation $z_1^n = z$ are given by

$$z_1 = r^{\frac{1}{n}}\left(\cos\left(\frac{(\theta + 2k\pi)}{n}\right) + i\sin\left(\frac{(\theta + 2k\pi)}{n}\right)\right) \quad \text{where } k = 0, 1, 2, \ldots, n-1$$

The position vectors of the solutions will divide the circle of radius r, centre the origin, into n equal sectors.

Example 1 Solve the equation $z^3 = 4 + i4\sqrt{3}$.

$|z^3| = \sqrt{(4^2 + (4\sqrt{3})^2)} = 8, \quad \arg(z^3) = \tan^{-1}\left(\frac{4\sqrt{3}}{4}\right) = \frac{\pi}{3}, \quad z^3 = 8\left(\cos\frac{\pi}{3} + i\sin\frac{\pi}{3}\right)$

Solutions are of the form

$$z = 8^{\frac{1}{3}}\left(\cos\frac{1}{3}\left(\frac{\pi}{3} + 2k\pi\right) + i\sin\frac{1}{3}\left(\frac{\pi}{3} + 2k\pi\right)\right) \quad \text{where } k = 0, 1, 2$$

$k = 0$ gives: $z = 2\left(\cos\frac{\pi}{9} + i\sin\frac{\pi}{9}\right)$

$k = 1$ gives: $z = 2\left(\cos\frac{7\pi}{9} + i\sin\frac{7\pi}{9}\right)$

$k = 2$ gives: $z = 2\left(\cos\frac{13\pi}{9} + i\sin\frac{13\pi}{9}\right)$

$= 2\left(\cos\left(-\frac{5\pi}{9}\right) + i\sin\left(-\frac{5\pi}{9}\right)\right)$

$= 2\left(\cos\frac{5\pi}{9} - i\sin\frac{5\pi}{9}\right)$

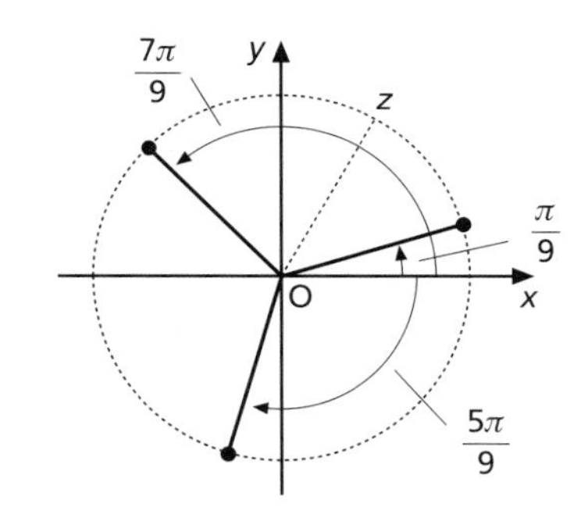

Example 2 Solve the equation $z^5 = 1$.

$|z^5| = 1, \quad \arg(z^5) = 0, \quad z^5 = 1(\cos 0 + i\sin 0)$

Solutions are of the form

$$z = 1^{\frac{1}{5}}\left(\cos\frac{1}{5}(0 + 2k\pi) + i\sin\frac{1}{5}(0 + 2k\pi)\right) \quad \text{where } k = 0, 1, 2, 3, 4$$

$k = 0$ gives: $z = (\cos 0 + i\sin 0) = 1$

$k = 1$ gives: $z = \left(\cos\frac{2\pi}{5} + i\sin\frac{2\pi}{5}\right)$

$k = 2$ gives: $z = \left(\cos\frac{4\pi}{5} + i\sin\frac{4\pi}{5}\right)$

$k = 3$ gives: $z = \left(\cos\frac{6\pi}{5} + i\sin\frac{6\pi}{5}\right) = \cos\left(-\frac{4\pi}{5}\right) + i\sin\left(-\frac{4\pi}{5}\right) = \cos\frac{4\pi}{5} - i\sin\frac{4\pi}{5}$

$k = 4$ gives: $z = \left(\cos\frac{8\pi}{5} + i\sin\frac{8\pi}{5}\right) = \cos\left(-\frac{2\pi}{5}\right) + i\sin\left(-\frac{2\pi}{5}\right) = \cos\frac{2\pi}{5} - i\sin\frac{2\pi}{5}$

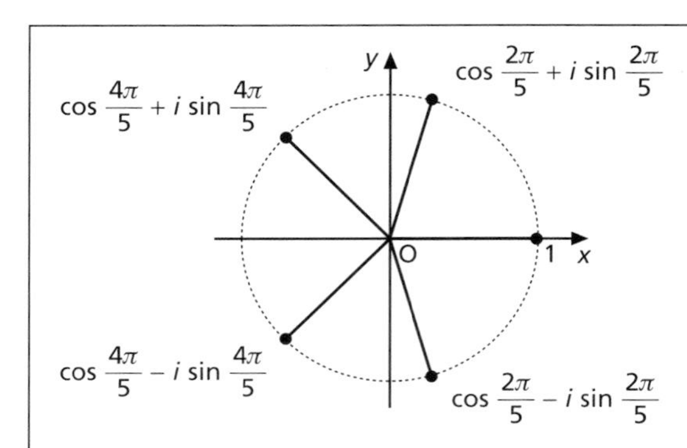

These solutions are often referred to as the *fifth roots of unity.*

When raised to the power of 5, each number produces the answer 1.

EXERCISE 7

1 For each of the following,

(i) solve the equation, leaving your answers in polar form,

(ii) draw an Argand diagram to illustrate the solutions.

a $z^3 = 8\left(\cos\frac{\pi}{4} + i\sin\frac{\pi}{4}\right)$

b $z^4 = \left(\cos\frac{\pi}{5} + i\sin\frac{\pi}{5}\right)$

c $z^5 = 32\left(\cos\frac{\pi}{7} + i\sin\frac{\pi}{7}\right)$

d $z^3 = 64\left(\cos\frac{2\pi}{3} + i\sin\frac{2\pi}{3}\right)$

e $z^5 = 32\left(\cos\frac{\pi}{7} - i\sin\frac{\pi}{7}\right)$

f $z^3 = 64\left(\cos\frac{2\pi}{3} - i\sin\frac{2\pi}{3}\right)$

g $z^4 = -2 - 2i$

h $z^5 = -3 + 3\sqrt{3}i$

i $z^4 = -2 + 2i$

j $z^5 = -3 - 3\sqrt{3}i$

2 **a** Solve $z^3 = 1$ to find the cube roots of unity. [Hint: $z^3 = 1(\cos 0 + i\sin 0)$.]

b Solve $z^4 = 1$ to find the fourth roots of unity.

c Find the sixth roots of unity.

d Solve the equation $z^4 = 81$.

e Find the complex numbers which satisfy the equation

(i) $z^5 = -1$ **(ii)** $z^5 = i$ **(iii)** $z^5 = -i$

f Find in polar form, the solutions of

(i) $z^3 = -64$ **(ii)** $z^4 = 625i$ **(iii)** $z^5 = \frac{-i}{32}$

Polynomials

Gauss

In 1799 Gauss proved that every polynomial equation with complex coefficents, $f(z) = 0$, where $z \in C$, has at least one root in the set of complex numbers.

He later called this theorem *the fundamental theorem of algebra.*

We restrict ourselves in this course to real coefficients.
The fundamental theorem of algebra still applies since real numbers are also complex.

If the root of a polynomial equation is non-real, i.e. of the form $r(\cos\theta + i\sin\theta)$, $r\sin\theta \neq 0$, then its conjugate, $r(\cos\theta - i\sin\theta)$, is also a root.

Proof

Suppose $F(z) = a_n z^n + a_{n-1}z^{n-1} + a_{n-2}z^{n-2} + \cdots + a_2 z^2 + a_1 z^1 + a_0$.

If $z = r(\cos\theta + i\sin\theta)$ is a root then

$$a_n r^n(\cos\theta + i\sin\theta)^n + a_{n-1}r^{n-1}(\cos\theta + i\sin\theta)^{n-1} + \cdots + a_1 r^1(\cos\theta + i\sin\theta)^1 + a_0 = 0$$

Using De Moivre's theorem and then equating real and imaginary parts:

$$a_n r^n \cos n\theta + a_{n-1}r^{n-1}\cos(n-1)\theta + \cdots + a_1 r^1 \cos\theta + a_0 = 0 \qquad (1)$$

$$a_n r^n \sin n\theta + a_{n-1}r^{n-1}\sin(n-1)\theta + \cdots + a_1 r^1 \sin\theta + a_0 = 0 \qquad (2)$$

Now

$$\begin{aligned}F(\bar{z}) &= a_n r^n(\cos\theta - i\sin\theta)^n + a_{n-1}r^{n-1}(\cos\theta - i\sin\theta)^{n-1} + \cdots \\ &\quad + a_1 r^1(\cos\theta - i\sin\theta)^1 + a_0 \\ &= a_n r^n(\cos n\theta - i\sin n\theta) + a_{n-1}r^{n-1}(\cos(n-1)\theta - i\sin(n-1)\theta) + \cdots \\ &\quad + a_1 r(\cos\theta - i\sin\theta) + a_0 \\ &= [a_n r^n \cos n\theta + a_{n-1}r^{n-1}\cos(n-1)\theta + \cdots + a_1 r^1\cos\theta + a_0] \\ &\quad - i\,[a_n r^n \sin n\theta + a_{n-1}r^{n-1}\sin(n-1)\theta + \cdots + a_1 r^1 \sin\theta + a_0] \\ &= 0 - i0 = 0\end{aligned}$$

Hence $\bar{z} = r(\cos\theta - i\sin\theta)$ is also a root.

A polynomial of degree n will have n complex roots.

Proof

Given $F(z) = 0$ is a polynomial equation of degree n then, by the fundamental theorem of algebra, a root exists. Call this root k_n. By the factor theorem $(z - k_n)$ is a factor and $F(z) = (z - k_n)G(z)$ where $G(z)$ is a polynomial of degree $n - 1$.
Since $F(z) = 0$ then $G(z) = 0$.

By the fundamental theorem of algebra a root for this polynomial equation also exists. Call this root k_{n-1}. By the factor theorem $(z - k_{n-1})$ is a factor and $F(z) = (z - k_n)(z - k_{n-1})H(z)$ where $H(z)$ is a polynomial of degree $n - 2$. Since $F(z) = 0$ then $H(z) = 0$.

Proceeding in this manner we get $F(z) = (z - k_n)(z - k_{n-1}) \ldots (z - k_2)(z - k_1)$.
Thus a polynomial of degree n will have n complex roots.

A polynomial of degree n, with real coefficients, can be reduced to a product of real linear factors and real irreducible quadratic factors.

Either a root, k, is real, leading to a linear factor $(z - k)$,
or it is non-real, leading to the non-real linear factors $(z - k)$ and $(z - \bar{k})$.

The product of these two factors leads to a real quadratic factor which is irreducible in R.

Example Given $f(z) = z^4 - 6z^3 + 18z^2 - 30z + 25$,

a show that $z = 1 + 2i$ is a root of the equation $f(z) = 0$,

b hence find the other roots of the polynomial.

a $f(1 + 2i) = (1 + 2i)^4 - 6(1 + 2i)^3 + 18(1 + 2i)^2 - 30(1 + 2i) + 25$

$= (-7 - 24i) - 6(-11 - 2i) + 18(-3 + 4i) - 30(1 + 2i) + 25$

$= -7 - 24i + 66 + 12i - 54 + 72i - 30 - 60i + 25$

$= 0$

Thus $z = 1 + 2i$ is a root.

b If $z = 1 + 2i$ is a root then, since $f(z)$ has real coefficients, the conjugate, $1 - 2i$, is also a root.

Thus $(z - (1 + 2i))$ and $(z - (1 - 2i))$ are complex factors of $f(z)$. These multiply to give the real quadratic factor $(z - 1 - 2i)(z - 1 + 2i) = z^2 - 2z + 5$.

By division

$$\begin{array}{r|l} & z^2 - 4z + 5 \\ \hline z^2 - 2z + 5 & z^4 - 6z^3 + 18z^2 - 30z + 25 \\ & z^4 - 2z^3 + 5z^2 \\ \hline & -4z^3 + 13z^2 - 30z + 25 \\ & -4z^3 + 8z^2 - 20z \\ \hline & 5z^2 - 10z + 25 \\ & 5z^2 - 10z + 25 \\ \hline & 0 \end{array}$$

We find that the complementary real factor is $z^2 - 4z + 5$.

Equating this to zero and using the quadratic formula

$$z = \frac{4 \pm \sqrt{(16 - 20)}}{2} = 2 \pm i$$

We now have all four roots, which are $1 + 2i$, $1 - 2i$, $2 + i$ and $2 - i$.

EXERCISE 8

1 Use the quadratic formula to find the two complex roots of each of the following equations.

a $z^2 - 2z + 10 = 0$ **b** $z^2 - 4z + 5 = 0$ **c** $z^2 - 6z + 25 = 0$

d $4z^2 - 16z + 17 = 0$ **e** $2z^2 + 2z + 1 = 0$ **f** $5z^2 + 4z + 8 = 0$

2 For each cubic equation, a real root has been identified. Find the remaining two complex roots.

a $z^3 + 2z^2 + z + 2 = 0;\quad z = -2$ **b** $z^3 - z^2 - z - 15 = 0;\quad z = 3$

c $z^3 + 6z^2 + 37z + 58 = 0;\quad z = -2$ **d** $z^3 - 11z - 20 = 0;\quad z = 4$

e $4z^3 + 8z^2 + 14z + 10 = 0;\quad z = -1$ **f** $2z^3 + 11z^2 + 20z - 13 = 0;\ z = \frac{1}{2}$

3 For each cubic equation, a factor has been given. Find the roots of the equation.

a $z^3 + z^2 - 7z - 15 = 0;\quad z - 3$

b $z^3 - 6z^2 + 13z - 20 = 0;\quad z - 4$

c $2z^3 + 5z^2 + 8z + 20 = 0;\quad z - 2i$

d $2z^3 + 7z^2 + 26z + 30 = 0;\quad z + 1 + 3i$

e $4z^3 - 4z^2 + 9z - 9 = 0;\quad 2z - 3i$

f $2z^3 + 13z^2 + 46z + 65 = 0;\quad z + 2 - 3i$

4 Given $f(z) = z^4 - z^3 - 2z^2 + 6z - 4$ and that $z = 1 + i$ is a root of the equation $f(z) = 0$, find the real factors of the polynomial.

5 $1 + 2i$ is a root of the equation $z^4 - 5z^3 + 13z^2 - 19z + 10$. Find the real factors of the polynomial.

6 In each of the following,

(i) show that the given complex number is a zero of the given polynomial,

(ii) find all the remaining roots.

a $z = 2 + i;\quad f(z) = z^4 - 2z^3 - z^2 + 2z + 10$

b $z = 3 + 2i;\quad f(z) = z^4 - 8z^3 + 30z^2 - 56z + 65$

c $z = 1 + 3i;\quad f(z) = 2z^4 - 3z^3 + 17z^2 + 12z - 10$

7 In each case the given complex number is a zero of the given polynomial. Find all the roots.

a $z = 4 + i;\quad f(z) = z^4 - 8z^3 + 13z^2 + 32z - 68$

b $z = 2 - 3i;\quad f(z) = 6z^4 - 31z^3 + 108z^2 - 99z + 26$

c $z = 4 + 2i;\quad f(z) = 2z^4 - 25z^3 + 107z^2 - 140z - 100$

8 $z^2 + 4z + 5$ is a factor of $z^5 + 7z^4 + 21z^3 + 33z^2 + 28z + 10$. Find all the roots of the equation $z^5 + 7z^4 + 21z^3 + 33z^2 + 28z + 10 = 0$.

Historical note

In the eighteenth century a Cambridge mathematician, Robert Cotes (1682–1716), developed a third form in which to express a complex number.

He discovered that

$$z = r(\cos\theta + i\sin\theta) = re^{i\theta}$$

where r is the modulus, θ is the argument of z and $e = 2.718\ldots$

Only by unit 3 of this course will you have enough information to prove this. It does, however, produce a few curiosities worthy of mention here:

When $z = -1$, $r = 1$ and $\theta = \pi$, which leads to a famous result: $\boxed{e^{i\pi} = -1}$

When $z = i$, $r = 1$ and $\theta = \frac{\pi}{2}$. $\quad i = e^{\frac{i\pi}{2}} \Rightarrow i^i = \left(e^{\frac{i\pi}{2}}\right)^i = e^{-\frac{\pi}{2}} \Rightarrow i^i \in R!!$

CHAPTER 4 REVIEW

1 Given $z_1 = 5 - 12i$ and $z_2 = 3 + 4i$, calculate:

a $z_1 + z_2$ **b** $z_1 - z_2$ **c** $z_1 \times z_2$ **d** $\overline{z}_1$

e $\dfrac{z_1}{z_2}$ **f** $\sqrt{z_1}$ **g** $2z_1 + z_2^2$

2 If $(2 + bi)(a + 3i) = 1 + 8i$, $a, b, \in N$, find a and b by equating real and imaginary parts.

3 Express $z = 5 + 12i$ in polar form.

4 Calculate the modulus and argument of $\dfrac{4 - 3i}{1 + i}$ correct to three significant figures where appropriate.

5 Draw an Argand diagram to illustrate the sum $z_1 + z_2$ where $z_1 = 1 + 3i$ and $z_2 = 4 + 2i$.

6 Given $z = x + iy$,

a find the equation of the locus $|z + 1| = 5$,

b draw the locus on an Argand diagram.

7 For $z_1 = 2\left(\cos\dfrac{\pi}{3} + i\sin\dfrac{\pi}{3}\right)$ and $z_2 = 3\left(\cos\dfrac{\pi}{6} + i\sin\dfrac{\pi}{6}\right)$ calculate:

a $z_1 z_2$ **b** $\dfrac{z_1}{z_2}$ **c** z_1^3

8 Use De Moivre's theorem to simplify $(-1 + \sqrt{3}i)^8$ giving your answer

a in polar form

b in the form $a + ib$

9 **a** Expand $(\cos\theta + i\sin\theta)^5$

(i) using the binomial theorem,

(ii) using De Moivre's theorem.

b By equating imaginary parts, express $\sin 5\theta$ in terms of $\sin\theta$.

10 **a** Solve $z^3 = 125\left(\cos\dfrac{\pi}{4} + i\sin\dfrac{\pi}{4}\right)$, leaving your answers in polar form.

b Illustrate the roots on an Argand diagram.

11 Find the fifth roots of unity and illustrate them on an Argand diagram.

12 Find the real factors of $f(z) = z^4 - z^3 + 13z^2 + 21z - 34$ given that $1 + 4i$ is a zero of $f(z)$.

13 Solve $z^4 - 6z^3 + 26z^2 - 46z + 65 = 0$ for all its complex roots given that $2 + 3i$ is one root.

CHAPTER 4 SUMMARY

1 **(i)** $i^2 = -1;\quad i \notin R$

(ii) C is the set of numbers z, of the form $z = a + bi$ where $a, b \in R$.

(iii) The members of C are called *complex numbers*.

(iv) a is called the *real* part of z and we write $a = \mathcal{R}(z)$.
b is called the *imaginary* part of z and we write $b = \mathcal{I}(z)$.

(v) Given $z_1 = a + bi$ and $z_2 = c + di$,
addition is defined by

$$z_1 + z_2 = (a + bi) + (c + di) = (a + c) + (b + d)i$$

multiplication is defined by

$$\begin{aligned} z_1 z_2 &= (a + bi)(c + di) = ac + adi + bci + bdi^2 \\ &= (ac - bd) + (ad + bc)i \qquad \text{since } i^2 = -1 \end{aligned}$$

(vi) If $z_1 = z_2$ then $\mathcal{R}(z_1) = \mathcal{R}(z_2)$ and $\mathcal{I}(z_1) = \mathcal{I}(z_2)$.

2 If $z = a + bi$ then its *conjugate* is $z = a - bi$.

3 Divisions are simplified by multiplying numerator and denominator by the conjugate of the denominator.

$$\frac{(a + bi)}{(c + di)} = \frac{(a + bi)(c - di)}{(c + di)(c - di)} = \frac{(a + bi)(c - di)}{(c^2 + d^2)}$$

4 **(i)** A complex number, z, can be represented on the complex plane by the point P or its position vector $\overrightarrow{OP}$.

(ii) Such a diagram is often referred to as an *Argand diagram*.

(iii) The angle through which OP has rotated is called the argument of z. Arg $z = \theta + 2n\pi$.

(iv) The principal argument lies in the range $(-\pi, \pi]$ and is denoted by arg z.

(v) The distance OP, r, is known as the modulus of z. This is denoted by $|z|$.

(vi) $r = \sqrt{(x^2 + y^2)}$ and $\theta = \tan^{-1}\left(\frac{y}{x}\right)$ remembering that one should also check in which quadrant P is to be found.

(vii) $x = r\cos\theta$ and $y = r\sin\theta$.

(viii) $z = r(\cos\theta + i\sin\theta)$ is the *polar form* of z.

5 **(i)** $|z_1 z_2| = |z_1| \times |z_2|$;
Arg $(z_1 z_2)$ = Arg (z_1) + Arg (z_2)

(ii) $\left|\frac{z_1}{z_2}\right| = |z_1| \div |z_2|$;
Arg $\left(\frac{z_1}{z_2}\right)$ = Arg (z_1) − Arg (z_2)

Note
The argument may have to be adjusted by adding or subtracting 2π to bring it back into the range $(-\pi, \pi]$.

6 De Moivre's theorem

$$z = r(\cos\theta + i\sin\theta) \Rightarrow z^n = r^n(\cos\theta + i\sin\theta)^n = r^n(\cos n\theta + i\sin n\theta)$$

7 **(i)** $z^n = r(\cos\theta + i\sin\theta) \Rightarrow z = r^{\frac{1}{n}}\left(\cos\left(\frac{(\theta + 2k\pi)}{n}\right) + \sin\left(\frac{(\theta + 2k\pi)}{n}\right)\right),$

$$k = 0, 1, 2, 3, \ldots, n - 1$$

(ii) As a special case, $z = 1$, $r = 1$ and $\theta = 0$, $z^n = 1$ has roots

$\left(\cos\left(\frac{2k\pi}{n}\right) + \sin\left(\frac{2k\pi}{n}\right)\right)$, $k = 0, 1, 2 \ldots, n - 1.$

These are the nth *roots of unity.*

8 **(i)** Every polynomial equation with complex coefficients, $f(z) = 0$, where $z \in C$, has at least one root in the set of complex numbers.
This is called *the fundamental theorem of algebra.*

(ii) If the root is non-real, i.e. of the form $r(\cos\theta + i\sin\theta)$, $r\sin\theta \neq 0$, then its conjugate, $r(\cos\theta - i\sin\theta)$, is also a root.

(iii) A polynomial of degree n will have n complex roots.

(iv) A polynomial of degree n, with real coefficients, can be reduced to a product of real linear factors and real irreducible quadratic factors.

5 Sequences and Series

Historical note

Babylonian tablet

Sequences and series have been studied in mathematics for as far back as written records go.

Geometric progressions have been deciphered on Babylonian tablets (2000 BC). Problems involving arithmetic and geometric sequences appear on the Ahmes Papyrus (1650 BC).

Euclid (c 300 BC) wrote 13 books referred to collectively as his *Elements*. In Book VIII he derives a formula for the sum of a geometric progression.

Recurrence relations

Reminders

- A *sequence* is an ordered list of numbers where each term is obtained according to a fixed rule.
- A *series*, or *progression*, is a sum, the terms of which form a sequence.
- The nth term of a sequence is often denoted by u_n, so that, for example, u_1 is the first term.
- A sequence can be defined by a recurrence relation where u_{n+1} is given as a function of lower/earlier terms.
- A first-order recurrence relation is where u_{n+1} is given as a function of u_n.
- A first-order linear recurrence relation is where $u_{n+1} = ru_n + d$, where r and d are constants.
- A sequence can be defined by a formula for u_n given as a function, $u_n = f(n)$.

Being given the first few terms of a sequence is not enough to identify the sequence.

Example 1 Identify the next term in the sequence 1, 2, 3, …

Possible answers include:

1 if $u_n = \frac{1}{2}(6 - 9n + 6n^2 - n^3)$

2 if $u_n = \frac{1}{3}(6 - 8n + 6n^2 - n^3)$

3 if $u_n = \frac{1}{6}(6 - 5n + 6n^2 - n^3)$

4 if $u_n = n$

5 *if* $u_n = \frac{1}{6}(n^3 - 6n^2 + 17n - 6)$

6 if $u_n = \frac{1}{3}(n^3 - 6n^2 + 14n - 6)$

7 if $u_n = \frac{1}{2}(n^3 - 6n^2 + 13n - 6)$

If, however, we also know that the sequence is generated by a *first-order linear recurrence relation* then only one of the above fits the description.

We know $u_{n+1} = ru_n + d$ and $u_1 = 1$, $u_2 = 2$ and $u_3 = 3$.

Since $u_1 = 1$ and $u_2 = 2$ then $2 = 1 \times r + d \Rightarrow 2 = r + d$.

Since $u_2 = 2$ and $u_3 = 3$ then $3 = 2 \times r + d \Rightarrow 3 = 2r + d$.

Subtracting we get $1 = r$, and back-substitution gives $d = 1$.

Thus $u_{n+1} = u_n + 1$.

When $u_n = 3$ then $u_{n+1} = 3 + 1 = 4$.

Example 2 Find the first-order linear recurrence relation when $u_3 = 7$, $u_4 = 15$ and $u_5 = 31$.

Since $u_3 = 7$ and $u_4 = 15$ then $15 = 7r + d$.

Since $u_4 = 15$ and $u_5 = 31$ then $31 = 15r + d$.

Subtracting we get $16 = 8r$, and so $r = 2$.

Back-substitution gives $d = 1$.

Thus $u_{n+1} = 2u_n + 1$.

(Note that $u_1 = 1$ and $u_2 = 3$.)

Reminder

Fixed points

Given the relation $u_{n+1} = 3u_n - 4$, then, if, for some value of n, $u_n = 3$, the sequence would proceed: …, 3, 5, 11, 29, …

However, if $u_n = 2$, we generate: …, 2, 2, 2, 2, …

When this repetition happens, u_n is referred to as a *fixed point*.

In this case, if any other value of u_n is used apart from 2, then the relation generates terms whose values *move away* or *diverge* from the value 2. $u_n = 2$ is an *unstable* fixed point.

Given the relation $u_{n+1} = 0.5u_n + 2$, then, if, for some value of n, $u_n = 4$, the sequence would proceed: ..., 4, 4, 4, 4, ...

If any other value of u_n is used apart from 4, then the relation generates terms whose values *move towards* or *converge on* the value 4. $u_n = 4$ is a *stable* fixed point, often referred to as the limit of the recurrence relation.

In general, for the relation $u_{n+1} = ru_n + d$, we have a fixed point when $u_{n+1} = u_n$. Solving these equations simultaneously we get:

$$u_n = ru_n + d \Rightarrow u_n = \frac{d}{1 - r}, \quad r \neq 1$$

Only when $|r| < 1$ is the fixed point stable.

EXERCISE 1

1 Each of the following shows the first three terms of a first-order linear recurrence relation. For each, find **(i)** the next term, **(ii)** an expression for u_{n+1} in terms of u_n.

a 2, 4, 6 **b** 2, 5, 8 **c** 15, 9, 3
d 3, 6, 12 **e** 32, 16, 8 **f** 1, 4, 10
g 3, 5, 9 **h** 1, 6, 26 **i** 61, 29, 13

2 A first-order linear recurrence relation is such that $u_1 = 1$, $u_2 = 5$, $u_4 = 53$ and $u_5 = 161$.
a Express u_{n+1} in terms of u_n.
b State the value of **(i)** u_3 and **(ii)** u_6.

3 You are given that $u_{n+1} = ru_n + d$ and $r, d \in Z$.
a Express **(i)** u_3 and **(ii)** u_4 in terms of r, d and u_1.
b Given $u_1 = 2$, $u_3 = 10$ and $u_4 = 28$, calculate the value of **(i)** r, **(ii)** d.

4 Find the fixed point of each of the following recurrence relations and say whether it is stable or unstable.

a $u_{n+1} = 3u_n + 4$ **b** $u_{n+1} = 0.6u_n + 8$ **c** $u_{n+1} = 2 - 3u_n$
d $u_{n+1} = 11 - 0.1u_n$ **e** $u_{n+1} = u_n - 1$ **f** $u_{n+1} = 0.9u_n - 1$

5 The volume of water in a lochan varies according to the formula $u_{n+1} = 0.8u_n + 6$ where u_n is the volume in 1 000 000 litres at the end of month n (Jan = 1).
a Calculate the volume of water at the start of the year (u_0) if there are 22 million litres at the end of January.
b If the climate is such that the model is suitable for long-range forecasts, what will be the long-term situation?

Arithmetic sequences

Definitions

If a sequence is generated so that, for all n,

$$u_{n+2} - u_{n+1} = u_{n+1} - u_n = d$$

then the sequence is known as an *arithmetic sequence*. The constant, d, is referred to as the *common difference*.

$u_{n+1} - u_n = d \Rightarrow u_{n+1} = u_n + d$, a first-order linear recurrence relation ($r = 1$, $d \neq 0$)

Traditionally u_1 is represented by the letter a: $u_1 = a$.

The nth term

$$\boxed{u_n = a + (n-1)d}$$

Proof

$$\begin{aligned} u_1 &= a \\ u_2 &= a + d \\ u_3 &= a + 2d \\ u_4 &= a + 3d \\ &\vdots \end{aligned}$$

The pattern of terms suggests that $u_n = a + (n-1)d$.
Suppose that this is true for $n = k$. Then

$$\begin{aligned} & u_k = a + (k-1)d \\ & u_{k+1} = u_k + d \qquad \text{by definition} \\ \Rightarrow\quad & u_{k+1} = a + (k-1)d + d \\ \Rightarrow\quad & u_{k+1} = a + (k-1+1)d \\ \Rightarrow\quad & u_{k+1} = a + ((k+1)-1)d \\ \Rightarrow\quad & \text{the formula is true for } n = k+1 \end{aligned}$$

We know it is true for $n = 1$. So, by induction, it is true for all $n \in N$.

Example 1 Find **a** the nth term, **b** the 10th term of the arithmetic sequence 6, 11, 16, …

a

$$\begin{aligned} & a = 6 \qquad \text{by inspection} \\ & d = 11 - 6 \text{ (or } 16 - 11) = 5 \\ & u_n = a + (n-1)d \\ \Rightarrow\quad & u_n = 6 + 5(n-1) \quad \Rightarrow \quad u_n = 1 + 5n \end{aligned}$$

b $u_{10} = 1 + 5 \times 10 = 51$

Example 2 Find the arithmetic sequence for which $u_3 = 9$ and $u_7 = 17$.

$$u_3 = 9 \Rightarrow a + (3-1)d = 9 \quad \Rightarrow a + 2d = 9$$
$$u_7 = 17 \Rightarrow a + (7-1)d = 17 \Rightarrow a + 6d = 17$$

Subtracting gives: $4d = 8 \quad \Rightarrow d = 2$

Substituting gives: $a + 4 = 9 \quad \Rightarrow a = 5$

So the sequence starts 5, 7, 9, 11, and has an nth term, $u_n = 5 + 2(n-1) = 2n + 3$.

Example 3 Given the arithmetic sequence 2, 8, 14, 20, …, for what value of n is $u_n = 62$?

$$a = 2,\ d = 6 \Rightarrow u_n = 2 + 6(n-1) = 6n - 4$$
$$u_n = 62 \Rightarrow 6n - 4 = 62$$
$$\Rightarrow n = 11$$

EXERCISE 2A

1 Identify a and d in each of the following arithmetic sequences.

a 3, 5, 7, … **b** 4, 5, 6, … **c** 3, 1, −1, …
d −2, 2, 6, … **e** −2, −5, −8, … **f** 0, 3, 6, …
g 2, 1.9, 1.8, … **h** $\frac{1}{12}, \frac{1}{6}, \frac{1}{4}, \ldots$ **i** $\frac{1}{8}, \frac{3}{16}, \frac{1}{4}, \ldots$

2 Find the nth term for each of these arithmetic sequences.

a 4, 7, 10, … **b** 8, 5, 2, … **c** 9, 5, 1, …
d −3, 2, 7, … **e** 17, 11, 5, … **f** $\frac{1}{2}, \frac{5}{8}, \frac{3}{4}, \ldots$
g $\frac{3}{10}, \frac{2}{5}, \frac{1}{2}, \ldots$ **h** $\frac{8}{9}, \frac{2}{3}, \frac{4}{9}, \ldots$ **i** −0.6, −1.3, −2, …

3 **a** Find the value of n when $a = -3$, $d = 2$ and $u_n = 15$.
b Which term in the arithmetic sequence 4, 1, −2, … is −14?
c If, in an arithmetic sequence, $u_1 = 1$ and $u_2 = 1.5$, which term has the value 31?

4 **a** Find the value of d when $a = 8$ and $u_{12} = 41$.
b An arithmetic sequence starts with 3. Its 100th term is −393. What is the common difference between terms?
c The first term of an arithmetic sequence is 1. List the first four terms if the 10th term is 4.

5 **a** For a particular arithmetic sequence, $u_{20} = 65$. If $d = 3$, find a.
b An arithmetic sequence with 12 as the common difference between terms has 231 for its 19th term. How does the sequence start?
c Counting *down* in twos, where do you start, to finish on 4 after 23 terms?

6 **a** Identify the arithmetic sequence in each case by quoting the first four terms.
(i) $u_5 = 24$ and $u_{10} = 49$ **(ii)** $u_8 = 11$ and $u_{15} = -3$ **(iii)** $u_9 = 7$ and $u_{17} = 13$
b An arithmetic sequence has 106 as its 14th term and 64 for its 8th term.
(i) Identify the sequence.
(ii) Which number between 150 and 160 is a term of the sequence?

7 An arithmetic sequence is defined by $u_n = 2n + 6$.
a Find analytically the value of x such that $2u_x = u_{3x}$.
b If a sequence is defined by $u_n = pn + q$ show that if $2u_x = u_{3x}$ then p is a factor of q.

8 By considering u_x and u_{x+1} show that a sequence defined by $u_n = pn + q$, where p and q are constants, is an arithmetic sequence.

9 a Show that ln 2, ln 6, ln 18, ... could be the first three terms of an arithmetic sequence and state the value of d, the common difference.

b Find u_n.

c For what value of x does u_x first exceed 100?

EXERCISE 2B

1 The fourth, fifth and sixth terms of an arithmetic sequence are x, $5x$ and $x + 8$ respectively.

a Calculate the value of x.

b What is the first term in the sequence?

c What is the first term bigger than 100?

2 A landscape gardener laying slabs for a patio puts four square slabs round a point.

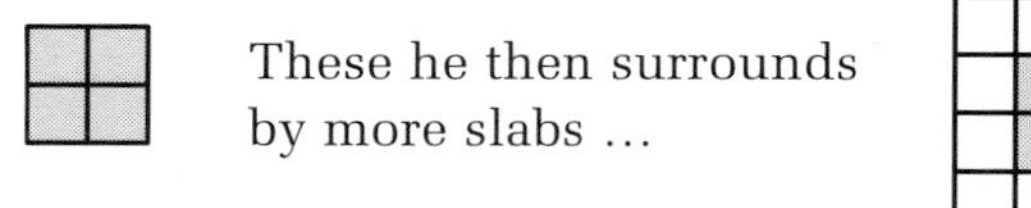

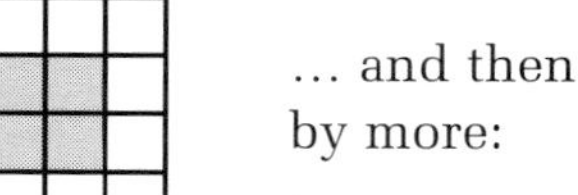

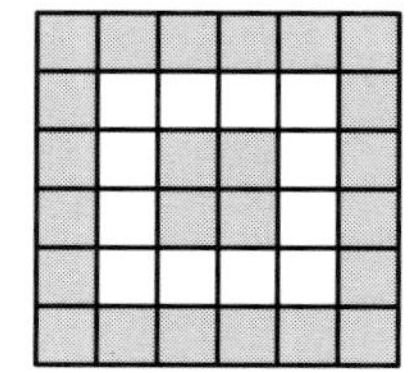

The numbers of slabs added at each stage form an arithmetic sequence.

a Identify the sequence by giving the first term and the common difference.

b As an alternative scheme, a single slab is used as the starting point.

(i) Verify that if the central stage is ignored then the additions at each of the other stages form an arithmetic sequence.

(ii) Give the first term and the common difference.

c If two slabs are used as the starting pattern, again they are part of the arithmetic sequence formed. Investigate when the starting stage is part of the sequence.

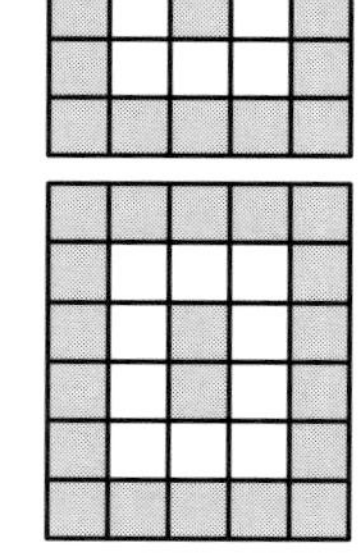

3 The numbers of seats in rows of a cinema form an arithmetic sequence. Counting from the front, the 10th row has 107 seats in it. The 20th row has 207 seats.

a How many seats are in the first row?

b What is the first row with more than 50 seats?

4 In a football stadium, the seats closest to the playing area number 700. The seats are built up in terraces which surround the pitch. The seats furthest from the action, in the 30th terrace, number 1280.

a If the numbers of seats in the terraces form an arithmetic progression, calculate the number of seats in the 10th terrace.

b In another stadium the front row seats 800 people and the 19th row seats 1000. Show that the numbers of seats in each row do not form an arithmetic sequence.

5 A kitchen roll comprises of a long unbroken length of tissue wrapped round a central hub. The tissue is 0.5 mm thick. Show that the circumferences of the complete turns of the tissue round the hub form an arithmetic sequence.
If the empty hub is 2 cm in diameter, work out the first term of the sequence and the common difference.

Historical note

According to Gauss, when he was 10 it was the habit of his teacher, Mr Büttner, to give the class a very long problem to solve, to keep them busy. He often gave them about 100 terms of an arithmetic sequence to add. This might be expected to occupy a 10 year old boy for about an hour. Gauss, however, totalled the sequence in seconds using the following ploy:

Consider that: $S = 1 + 2 + 3 + \cdots + 997 + 998 + 999$

Reverse the list: $S = 999 + 998 + 997 + \cdots + 3 + 2 + 1$

Add the two lists: $2S = 1000 + 1000 + 1000 + \cdots + 1000 + 1000 + 1000$

A simple multiplication: $2S = 999\,000$

Half the answer: $S = 499\,500$

The sum to n terms of an arithmetic sequence

An arithmetic series, or arithmetic progression (AP) is a sum whose terms form an arithmetic sequence. Its value, to n terms, denoted by S_n, can be found using:

$$S_n = \frac{1}{2}n(2a + (n-1)d)$$

Proof

$$S_n = a + (a + d) + (a + 2d) + (a + 3d) + \cdots + (a + (n-2)d) + (a + (n-1)d)$$

Reverse the order:

$$S_n = (a + (n-1)d) + (a + (n-2)d) + \cdots + (a + 3d) + (a + 2d) + (a + d) + a$$

Add: $2S_n = [(a + (n-1)d) + a] + [(a + (n-2)d) + (a + d)] + \cdots + [(a + d) + (a + (n-2)d)] + [a + (a + (n-1)d)]$

$$2S_n = (2a + (n-1)d) + (2a + (n-1)d) + \cdots + (2a + (n-1)d) + (2a + (n-1)d)$$

giving: $2S_n = n(2a + (n-1)d)$

Hence: $S_n = \frac{1}{2}n(2a + (n-1)d)$

Example 1 Find the sum of the first 15 terms of the arithmetic sequence which starts 3, 8, 13, 18, …

$a = 3$, $d = 5$, $n = 15$ so, since $S_n = \frac{1}{2}n(2a + (n-1)d)$, we have

$$S_{15} = \frac{15}{2}((2 \times 3) + (15 - 1) \times 5)) = \frac{15}{2}(6 + 70) = 570$$

Example 2 When does the sum of the arithmetic sequence which starts 2, 10, 18, 26, first exceed 300?

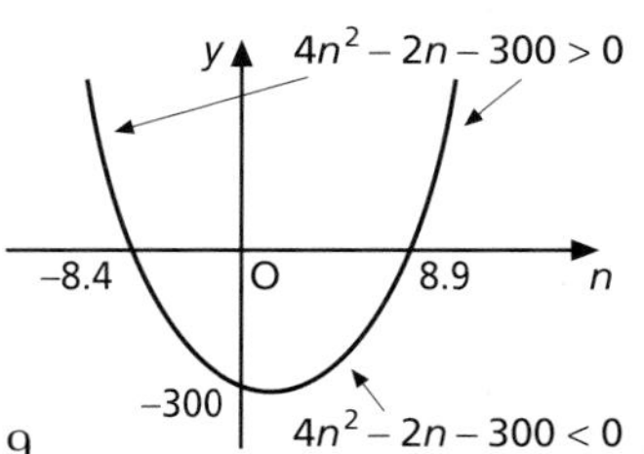

$a = 2$, $d = 8$, so $S_n = \frac{1}{2}n(4 + 8(n - 1)) = 4n^2 - 2n$

$$S_n > 300 \Rightarrow 4n^2 - 2n > 300 \Rightarrow 4n^2 - 2n - 300 > 0$$

Solving $4n^2 - 2n - 300 = 0$ gives

$n = -8.4$ or $n = 8.9$ (to 1 d.p.)

Given that $n \in N$, and from a quick sketch, we see that $n \geq 9$.

Example 3 The sum of the first four terms of an arithmetic sequence is 26. The sum of the first 12 terms is 222. What is the sum of the first 20 terms?

$$S_4 = 26 \Rightarrow 2(2a + 3d) = 26 \quad \Rightarrow 2a + 3d = 13$$
$$S_{12} = 222 \Rightarrow 6(2a + 11d) = 222 \quad \Rightarrow 2a + 11d = 37$$

Subtracting gives $8d = 24 \quad \Rightarrow d = 3$

Substituting gives $a = 2$

$$S_{20} = 10(4 + (19 \times 3)) = 610$$

Note that, given u_n, a simplified formula for the sum to n terms is

$$S_n = \frac{1}{2}n(a + u_n)$$

EXERCISE 3A

1 **a** Calculate the sum to 10 terms of the arithmetic series which starts $3 + 10 + 17 + \cdots$

b Find S_{12} for an arithmetic series when $u_1 = 8$, $u_2 = 27$ and $u_3 = 46$.

c Find the required sum when each of the following is an arithmetic series.

(i) $3 + 5 + 7 + \cdots$: S_{25} **(ii)** $7 + 11 + 15 + \cdots$: S_{100}

(iii) $(-2) + (-10) + (-18) + \cdots$: S_{16} **(iv)** $-5 - 3 - 1 - \cdots$: S_7

2 The first two terms of an arithmetic sequence are 9 and 12 in that order.

a Find the sum of the first **(i)** 18 terms, **(ii)** 19 terms.

b Hence calculate the 19th term.

c Repeat this process if the first two terms are 12 and 9 in that order.

3 Find the following sums, given that each is an arithmetic series.

a $2 + 3 + 4 + \cdots + 25$ **b** $12 + 19 + 26 + \cdots + 355$

c $-5 + (-7) + \cdots + (-43)$ **d** $8 - 2 - 12 - \cdots - 62$

e $0.01 + 0.32 + 0.63 + \cdots + 2.8$ **f** $\frac{1}{4} + \frac{1}{2} + \frac{3}{4} + \cdots + 4$

4 **a** The sum of the first 50 terms of an arithmetic series is 2800. The common difference is 2. What is the first term?

b The sum of the first 10 terms of an arithmetic series is 25. The common difference is 0.3. What is the first term?

5 **a** The first term of an arithmetic progression is 6 and the last term is 78. The sum of the terms is 798. How many terms are in the progression?

b List the first four terms of an arithmetic progression if the last term is ten times the first term and the sum of the terms is 121. [There are two possible solutions.]

6 **a** An arithmetic series starts with 3, has 25 terms and totals 2175. What is the common difference?

b What is the common difference of an arithmetic series which has 100 terms, the first term being 0.5, and which totals 545?

7 **a** The first three terms of an arithmetic sequence total 24. The next three total 69. What is the sum of the three after that?

b The sum of the first four terms of an arithmetic sequence is −36. The next five total −135. What is the 21st term?

8 **a** Given that $u_{14} = 20$ and $S_{15} = 165$, calculate u_{16} and S_{17}.

b The first term of a sequence is 10. The sum to 22 terms equals the 22nd term.
(i) What is the common difference? (ii) What is the 22nd term?

9 A historical problem (sixteenth century).
100 eggs are placed in a straight line, one step apart. A basket is situated one step from the start of the line. A person walks away from the basket, picks the first egg up and returns it to the basket. He then walks to the second egg, picks it up and returns it to the basket. He continues in this fashion until all the eggs are in the basket. How many steps did he take?

EXERCISE 3B

1 The first three terms of an arithmetic sequence are x, $x + 8$ and bx where $x, b \in N$.

a Express x in terms of b.

b Identify all the sequences which fit the description.

c Calculate the sum to 10 terms in each case.

2 An arithmetic sequence is such that $u_1 = x$ and $u_4 = x^3$.

a Prove that if $x \in N$, $x > 1$, then $d \in N$.

b (i) Find the sum to n terms when $x = 2, 3, 4, 5$ in terms of n.
(ii) Comment on the pattern created by the coefficients of n^2 in the context of Pascal's triangle.

3 A roll of tape is wound round a spindle of radius 2 cm. The tape is only 0.01 cm thick.

a Taking each complete winding as approximately circular, and keeping π in your answer, find the circumference of each of the first five windings.

b If there are 100 windings on the roll calculate the total length of tape on the roll.

4 An indoor sports arena has seating all the way round the running track. There are 300 seats in the tier closest to the track.

a If there are 15 tiers and each tier holds 25 more seats than the previous one, what is the capacity of the arena?

b For security reasons in one event, the front five tiers are to be left empty. At £34 a seat, what will be the takings if the event is a sell-out?

5 A bee builds a series of hexagonal cells around a central cell. The successive rings of cells form an arithmetic sequence (the central cell is not part of this sequence).

a How many rings are needed before the total number of cells exceeds 500?

b As an alternative, two cells can form the nucleus of the structure. How many layers are needed now before a total of 500 is reached?

c Investigate the problem for different sized nuclei.

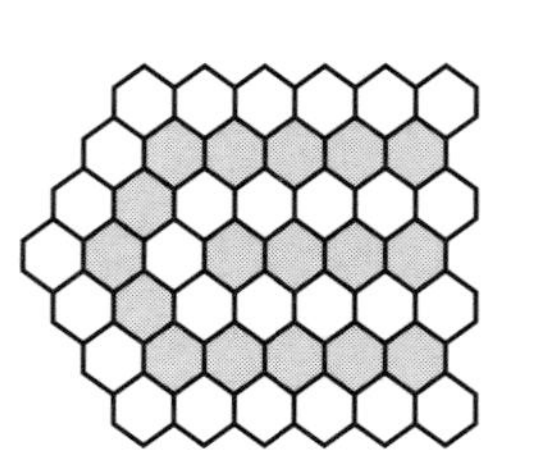

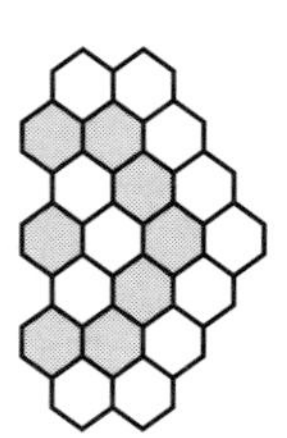

Geometric sequences

Definitions

If a sequence is generated so that, for all $n \in N$,

$$u_{n+2} \div u_{n+1} = u_{n+1} \div u_n = r$$

then the sequence is known as a geometric sequence. The constant, r, is referred to as the *common ratio.*

$u_{n+1} \div u_n = r \Rightarrow u_{n+1} = ru_n$ a first-order linear recurrence relation ($d = 0$, $r \neq 0, 1$)

Again u_1 is traditionally represented by the letter a: $u_1 = a$.

The nth term

$$u_n = ar^{(n-1)}$$

Proof

$$u_1 = a$$
$$u_2 = ar$$
$$u_3 = ar^2$$
$$u_4 = ar^3$$
$$\vdots$$

The pattern of terms suggests that $u_n = ar^{n-1}$. Suppose that this is true for $n = k$. Then

$$
\begin{aligned}
& u_k = ar^{k-1} \\
& u_{k+1} = u_k \times r \qquad \text{by definition} \\
\Rightarrow \quad & u_{k+1} = ar^{k-1} \times r \\
\Rightarrow \quad & u_{k+1} = ar^{k-1+1} \\
\Rightarrow \quad & u_{k+1} = ar^{(k+1)-1} \\
\Rightarrow \quad & \text{the formula is true for } n = k + 1
\end{aligned}
$$

We know it is true for $n = 1$. So, by induction, it is true for all $n \in N$.

Example 1 Find **a** the nth term, **b** the 10th term of the geometric sequence 3, 12, 48, ...

a

$$
\begin{aligned}
a &= 3 \qquad \text{by inspection} \\
r &= 12 \div 3 = 4 \\
u_n &= ar^{n-1} \\
\Rightarrow \quad u_n &= 3 \times 4^{(n-1)}
\end{aligned}
$$

b $u_{10} = 3 \times 4^{(10-1)} = 3 \times 4^9 = 786\,432$

Example 2 Find the geometric sequence whose third term is 18 and whose eighth term is 4374.

$$
\begin{aligned}
u_3 &= ar^2 = 18 \\
u_8 &= ar^7 = 4374
\end{aligned}
$$

Dividing gives: $r^5 = 243 \Rightarrow r = \sqrt[5]{243} = 3$

Substituting gives: $a \times 3^2 = 18 \Rightarrow a = 2$

So the sequence starts 2, 6, 18, 54, and has an nth term, $u_n = 2 \times 3^{(n-1)}$.

Example 3 Given the geometric sequence 5, 10, 20, 40, ..., find the value of n for which $u_n = 20\,480$.

$$
\begin{aligned}
a = 5,\ r = 2 \quad \Rightarrow \quad & 5 \times 2^{n-1} = 20\,480 \\
\Rightarrow \quad & 2^{n-1} = 4096 \\
\Rightarrow \quad & \ln 2^{n-1} = \ln 4096 \\
\Rightarrow \quad & (n-1) \ln 2 = \ln 4096 \\
\Rightarrow \quad & n - 1 = \ln 4096 / \ln 2 = 12 \\
\Rightarrow \quad & n = 13
\end{aligned}
$$

EXERCISE 4A

1 For each of the following geometric sequences, **(i)** identify a and r, **(ii)** hence find an expression for the nth term.

a 1, 3, 9, 27, ...
b 4, –8, 16, –32, ...
c 1458, 486, 162, 54, ...
d 3072, –768, 192, –48, ...
e 7, 0.7, 0.07, 0.007, ...
f $\frac{2}{3}, \frac{4}{15}, \frac{8}{75}, \frac{16}{375}, \ldots$
g 0.16, 0.128, 0.1024, ...
h 23.2, 4.64, 0.928, ...

2 **a** The first term of a geometric sequence is 2 and the common ratio is 6. What is the fifth term?

b If the first term of a geometric sequence is 0.8 and the second term is 0.24, calculate the sixth term.

3 The sixth term of a geometric sequence is 31 250 and the tenth term is 19 531 250.

a Find an expression for the nth term.

b What is the difference between the eighth and ninth term?

4 **a** The first two terms of a geometric sequence are 1024 and 2560 in that order. Which term has a value of 625 000?

b If the initial two terms had been $u_1 = 2560$ and $u_2 = 1024$, determine the first term that has a value less than 1.

5 **a** The first term of a geometric sequence is 25. The tenth term is 139. Calculate the common ratio to two decimal places.

b The common ratio of a geometric sequence is 0.98. The eighth term is 131. What is the first term to the nearest whole number?

6 **a** Show that a, $a + d$, $a + 2d$ are not the first three terms of a geometric sequence.

b If a, $a + d$, $a + 6d$ are the first three terms of a geometric sequence, **(i)** express a in terms of d, **(ii)** find the common ratio.

c If a, $a + d$, $a + xd$ are the first three terms of a geometric sequence express the common ratio in terms of x only.

7 **a** Show that $2 \sin x$, $\sin 2x$ and $\sin 2x \cos x$ could be the first three terms of a geometric sequence.

b If $x = \frac{\pi}{4}$ radians express the first ten terms in their simplest form.

8 **a** If p, x, q are the first three terms of a geometric sequence then x is called the geometric mean between p and q.
Find the geometric mean between **(i)** 6 and 24 **(ii)** $\frac{3}{4}$ and $\frac{27}{25}$.

b 1250, x, y, 10 form a geometric sequence. Calculate x and y.

EXERCISE 4B

1 The gears on a bike work best when the number of teeth on the gear-wheels form a geometric sequence. The terms, of course, must be rounded to the nearest whole number.

a Calculate the unknown number of teeth in each of these three-wheel gears.

(i) 12 teeth, x teeth, 24 teeth **(ii)** 15 teeth, 20 teeth, y teeth

(iii) z teeth, 20 teeth, 26 teeth

b A certain type of gear has four wheels. Again, working to the nearest whole number, calculate the unknown terms.

(i) 12 teeth, x teeth, y teeth, 36 teeth **(ii)** 15 teeth, 20 teeth, p teeth, q teeth

(iii) a teeth, b teeth, 25 teeth, 30 teeth **(iv)** c teeth, 20 teeth, d teeth, 26 teeth

2 The population of a particular endangered species increases annually by 3%. Initially there are 2000 animals.

a Calculate the population after **(i)** one year **(ii)** two years **(iii)** three years

b Show that the successive populations form a geometric sequence and state the common ratio.

c The species will be taken off the danger list when its population exceeds 5000. After how many years will this happen?

3 A bank offers compound interest at a rate of $r\%$ per year. Initially £P is deposited in the bank.

a Find an expression for A_1, the amount of money (in £) in the bank after one year.

b Find an expression for A_2, in terms of P and r.

c The successive amounts in the bank form a geometric sequence. Find an expression for the common ratio in terms of r.

d Hence or otherwise find an expression for A_n, the amount of money in the bank after n years.

4 In an experiment, a pendulum was released and allowed to swing. Its first swing covered an arc of 10°. Its second swing covered an arc of 9.5°.

a If successive swings form a geometric sequence, calculate the size of arc swept in the sixth swing.

b The pendulum is effectively stilled when the arc swept is less than 1°. How many swings will the pendulum make from release to when it is stilled?

5 A virus spreads in a community according to the model $N_t = N_0 e^{at}$, where N_0 is the number of people with the virus at the start and N_t is the number of people with the virus after t days.

a **(i)** Show that, when the number of people with the virus each day is listed, it forms a geometric sequence.

(ii) State the common ratio.

b If $N_0 = 2$ and $N_1 = 7$, when will the number of people with the virus first exceed 1000?

6 The value of a car depreciates according to the law $V_t = V_0 e^{-\frac{t}{10}}$, where V_0 is the value of the car when new and V_t is the value after t years.

a **(i)** Show that successive values taken over any regular period of time, T, form a geometric sequence.

(ii) State the common ratio in terms of T.

b Over what period of time will

(i) the original value be halved,

(ii) the original value be 10% of the original?

The sum to n terms of a geometric series

A geometric series, or geometric progression (GP), is a sum whose terms form a geometric sequence. Its value, to n terms, can be found using:

$$S_n = \frac{a(1 - r^n)}{1 - r}$$

Proof

We wish to find $\quad S_n = a + ar + ar^2 + ar^3 + \cdots + ar^{n-2} + ar^{n-1}$

Multiply throughout by r: $\quad rS_n = \quad ar + ar^2 + ar^3 + \cdots + ar^{n-2} + ar^{n-1} + ar^n$

Subtracting, we get: $\quad S_n - rS_n = a - ar^n$

$$\Rightarrow (1 - r)S_n = a(1 - r^n)$$

$$\Rightarrow \quad S_n = \frac{a(1 - r^n)}{1 - r} \qquad (r \neq 1)$$

Example 1 Find the sum to six terms of the geometric sequence whose first term is 4 and whose common ratio is 1.5.

$a = 4, \quad r = 1.5, \quad n = 6$

$$\Rightarrow \quad S_6 = \frac{4(1 - 1.5^6)}{1 - 1.5}$$

$$= 83.125$$

Example 2 A geometric sequence starts 12, 15, 18.75, … What is the smallest value of n for which S_n is bigger than 100?

$a = 12, \quad r = 1.25$

We require $\dfrac{12(1 - 1.25^n)}{1 - 1.25} > 100$

$$\Rightarrow \quad \frac{12(1.25^n - 1)}{0.25} > 100$$

$$\Rightarrow \quad 1.25^n - 1 > 100 \times (0.25) \div 12$$

$$\Rightarrow \quad 1.25^n > \frac{25}{12} + 1$$

$$\Rightarrow \quad 1.25^n > \frac{37}{12}$$

$$\Rightarrow \quad n \ln 1.25 > \ln\left(\frac{37}{12}\right)$$

$$\Rightarrow \quad n > \frac{\ln\left(\frac{37}{12}\right)}{\ln 1.25}$$

$$\Rightarrow \quad n > 5.046 \quad \text{(3 d.p.)}$$

$n \in N$ so the smallest value of n is 6

Example 3 A geometric series is such that $S_3 = 14$ and $S_6 = 126$. Identify the series.

$$14 = \frac{a(1 - r^3)}{1 - r} \quad \text{and} \quad 126 = \frac{a(1 - r^6)}{1 - r} \quad (r \neq 1)$$

Dividing we get: $$9 = \frac{(1 - r^6)}{(1 - r^3)}$$

$$\Rightarrow r^6 - 9r^3 + 8 = 0$$

Solving as a quadratic in r^3 gives

$$r^3 = 8 \text{ or } 1$$

$$\Rightarrow \quad r = 2 \quad (r \neq 1)$$

also $14 = \dfrac{a(1 - 2^3)}{1 - 2} \Rightarrow a = 2$

So the series is $2 + 4 + 8 + \cdots$

EXERCISE 5A

1 Sum each of the following geometric sequences to the required number of terms.

a 2, 6, 18, ... to 7 terms
b 4, 8, 16, ... to 9 terms
c 3, −15, 75, ... to 8 terms
d 4, 20, 100, ... to 6 terms
e 1, −2, 4, ... to 10 terms
f 2, −8, 32, ... to 7 terms

2 Calculate each of these geometric series to the required number of terms.

a $\frac{1}{4} + \frac{1}{12} + \frac{1}{36} + \cdots$ to 6 terms
b $\frac{4}{3} + \frac{1}{2} + \frac{3}{16} + \cdots$ to 8 terms
c $12 - 10 + \frac{25}{3} - \cdots$ to 7 terms
d $\frac{1}{4} + \frac{2}{5} + \frac{16}{25} + \cdots$ to 9 terms
e $4 - \frac{1}{3} + \frac{1}{36} - \cdots$ to 10 terms
f $50 - 20 + 8 - \cdots$ to 8 terms

3 a How many terms must be added for the geometric series $9 + 18 + 36 + \cdots$ to equal 9207?

b How many terms must be added for the geometric series $9 + 6 + 4 + \cdots$ to exceed 26?

4 A geometric series has a common ratio of 2. Its sum to 6 terms is 42.

a Calculate the first term.

b Find the sum to eight terms.

5 a For the geometric series with a first term a and a common ratio r show that

$$u_n + u_{n+1} = r^{n-1}(u_1 + u_2)$$

b The first two terms of a geometric series add up to 4. The fourth and fifth terms total 108. Identify the series.

c The first three terms of a geometric series total 7. The sixth, seventh and eighth total 224. List the first five terms of the series.

d The first three terms of a geometric series total 93. The first six terms of the series total 11 718.

(i) State the sum of the fourth, fifth and sixth terms.

(ii) Identify the series.

(iii) Compare this method with that of Example 3.

6 x, $x + 1$, $x + 4$ are the first three terms of a geometric sequence.
  a What is the value of x?
  b Calculate the sum of the first ten terms.

7 Edouard Lucas, a nineteenth-century French mathematician, studied sequences which are now named after him. A Lucas sequence is one where $u_{n+1} = u_n + u_{n-1}$, i.e. each term is the sum of the previous two terms in the sequence.
  a Letting $u_1 = 1$ and $u_2 = x$, find a Lucas sequence which is also a geometric sequence.
  b Let ϕ represent the common ratio of this sequence.
    (i) Find an expression for S_n in terms of ϕ.
    (ii) Verify for $n = 6$, $n = 7$ and $n = 8$ that S_n can also be expressed as $\phi^{n+1} - \phi$.
  c It is said that any Lucas sequence behaves more and more like a geometric sequence as the size of n increases. Investigate this claim.

8 a For each of the following sequences, calculate, where practical, S_{10}, S_{20}, S_{30}.
    (i) $a = 1, r = 2$ (ii) $a = 1, r = 0.2$ (iii) $a = 2, r = -2$
    (iv) $a = 2, r = -0.4$ (v) $a = 20, r = 0.9$ (vi) $a = 1000, r = -0.7$
  b Comment on your results when $|r| < 1$.

EXERCISE 5B

1 When making a guitar, the spacings between the frets on the neck are mathematically fixed. Each spacing is $\frac{17}{18}$ of the previous spacing.
  a If the first spacing (between the nut and first fret) is 3 cm, calculate the spacing between the fifth and seventh fret.
  b If the twelfth fret is placed half-way between the nut and the bridge, what is the distance between the nut and the bridge?

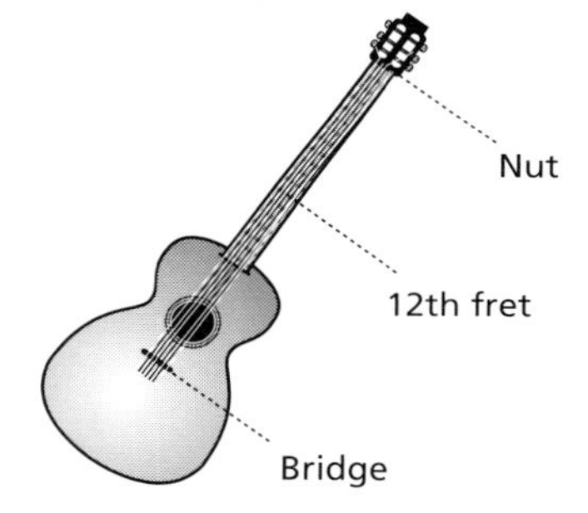

2 A building project, planned over five years, will cost in total £1 000 000. This estimate is based on a 3% rate of inflation, i.e. the costs will go up 3% each year.
  a Calculate the estimated costs for each of the five years.
  b If, instead, the actual rate of inflation is 4%, calculate the extra cost of the five-year project.

3 A pendulum is slowly coming to rest. With each swing, the arc it sweeps is reduced by 2%. Each swing takes the same time, 1.2 seconds. The first sweep is a 12° arc.
  a How many degrees has the pendulum swept through in five swings?
  b How long will it take before a total exceeding 90° has been swept through?

4 A certain type of tree used for hedging is planted when it is 1 metre tall. It grows so that its annual increase in height in any year is $\frac{11}{12}$ of the increase in height in the previous year. In its first year it grew by 1 metre.
  a What is the height of the plant after (i) 1 year, (ii) 2 years, (iii) 7 years?
  b (i) What is the height after 15 years?
    (ii) Compare the heights after 7 years and after 15 years.

5 On a tour of a Highland distillery the guides like to tell about the *Angels' share.* When whisky is distilled it is stored in large wooden casks for eight years to mature. Each year, because of evaporation, a certain amount is lost. These successive losses form a geometric sequence whose sum after eight years represents 10% of the volume of the cask when originally laid down. This loss is referred to as the Angels' share.

a If V_0 represents the original volume, find an expression for V_n, the volume at the end of year n when $n = 1, 2, 3$ and 4 (work to 2 s.f.).

b A deluxe whisky is laid down for 12 years. What is the Angels' share in this case?

6 A first-order linear recurrence relation is defined by $u_{n+1} = ru_n + d$.

a If $u_1 = a$, write down expressions in a, d and r for:

(i) u_1 **(ii)** u_2 **(iii)** u_3 **(iv)** u_4

b Use the formula for a sum of a geometric series to find an expression for u_n.

c Investigate the possibility of getting an expression for the sum to n terms of the recurrence relation.

Infinite series, partial sums, sum to infinity

An *infinite series* is a series which has an infinite number of terms.

When we have an infinite series then S_n is defined as the sum of the first n terms of that series. Such a sum is referred to as a *partial sum* of the series.

If the partial sum, S_n, tends towards a limit as n tends to infinity, then the limit is called the *sum to infinity* of the series.

Arithmetic series

Consider the sum to n terms of an arithmetic series.

$$S_n = \frac{n}{2}(2a + (n - 1)d)$$

This can be rearranged as

$$S_n = n^2\left[\frac{d}{2} + \frac{\left(a - \frac{d}{2}\right)}{n}\right]$$

For large values of n, $\dfrac{\left(a - \frac{d}{2}\right)}{n}$ becomes negligible and $S = \dfrac{d}{2}n^2$ becomes a good approximation for the sum.

As $n \to \infty$ we see that the sum $S_\infty \to \pm\infty$ depending on the value of d.

The sum to infinity for an arithmetic series is undefined.

Geometric series

Consider the sum to n terms of a geometric series.

$$S_n = \frac{a(1-r^n)}{1-r}, \qquad r \neq 1$$

To examine what happens for large values of n we need only note the behaviour of r^n.

When $|r| > 1$ then $r^n \to \infty$ as $n \to \infty$.

When $|r| > 1$, the sum to infinity of a geometric series is undefined.

However, if $|r| < 1$ then $r^n \to 0$ as $n \to \infty$.

When $|r| < 1$, the sum to infinity of a geometric series is defined as $S_\infty = \dfrac{a}{1-r}$

Example 1 Find the sum to infinity of the geometric series 24 + 12 + 6, … if it exists.

$a = 24$; $r = 12 \div 24 = 0.5 \Rightarrow |r| < 1 \Rightarrow S_\infty$ exists

$$S_\infty = \frac{24}{1-0.5} = 48$$

Example 2 Express the recurring decimal 0.121 212 … as a vulgar fraction.

$0.121\,212\ldots = 0.12 + 0.001\,2 + 0.000\,012 + \cdots$

$a = 0.12$; $r = 0.01 \Rightarrow |r| < 1 \Rightarrow S_\infty$ exists

$$S_\infty = \frac{0.12}{1-0.01} = \frac{0.12}{0.99} = \frac{12}{99} = \frac{4}{33}$$

Example 3 Given that 12 and 3 are two adjacent terms of an infinite geometric progression with a sum to infinity of 64, find the first term.

Since a sum to infinity exists, $|r| < 1$. So $r = \frac{3}{12} = 0.25$.

$$64 = \frac{a}{1-0.25}$$

$$\Rightarrow \quad a = 48$$

EXERCISE 6A

1 Find the sum to infinity of the following infinite geometric progressions.

a $1 + 0.5 + 0.25 + \cdots$ **b** $12 + 6 + 3 + \cdots$ **c** $7 + 1 + \frac{1}{7} + \cdots$

d $8 - 4 + 2 - \cdots$ **e** $81 - 9 + 1 - \cdots$ **f** $100 + 10 + 1 + \cdots$

2 Identify which of the following geometric series tend to a limit and find the limit.

a $18 + 3.6 + 0.72 + \cdots$ **b** $0.5 - 1 + 2 - \cdots$ **c** $-5 - 2.5 - 1.25 - \cdots$

d $0.1 - 0.2 + 0.4 - \cdots$ **e** $\frac{3}{5} + \frac{6}{15} + \frac{12}{45} \cdots$ **f** $\frac{1}{12} + \frac{1}{6} + \frac{1}{3} \cdots$

3 a A geometric series has a sum to infinity of 40. If the common ratio is 0.2, what is the first term of the series?

b The first term of an infinite geometric series is 18. The third term is 2.88.

(i) Show that the series has a limit.

(ii) Find the limit.

4 Express each of the following recurring decimals as an infinite geometric series, and hence as a vulgar fraction in its simplest form.

a 0.141 41 ... **b** 0.144 44 ... **c** 0.270 270 ...

5 a By considering 0.014 14 ... as $\frac{1}{10}$(0.141 414 ...) express it as a vulgar fraction in its simplest form.

b By considering 0.614 14 ... as $\frac{6}{10} + \frac{1}{10}$(0.141 414 ...) express it as a vulgar fraction in its simplest form.

c Express each of the following recurring decimals as a vulgar fraction in its simplest form.

(i) 0.011 11 ... **(ii)** 0.322 222 ... **(iii)** 0.436 363 6 ... **(iv)** 0.501 818 181 81 ...

6 Given that 0.64 and 0.128 are two adjacent terms of an infinite geometric series with a sum to infinity of 20,

a find the first term,

b find the partial sum S_5.

7 A ball is thrown to the ground. It bounces to a height of 3 m. The characteristics of the rubber are such that, thereafter, the ball rebounds to 0.4 of its previous height at each bounce. Let u_n represent the distance travelled, from ground to ground, in the nth bounce ($u_1 = 6$ metres).

a Write down the first four terms of the geometric series generated.

b Work out the total distance travelled after

(i) five bounces,

(ii) ten bounces.

c What is the limit of this distance as n tends to infinity?

EXERCISE 6B

1 In the fifth century BC a Greek philosopher, Zeno of Elea, posed a series of paradoxes which puzzled his contemporaries. The most famous of these is his story of Achilles and the tortoise.

Achilles can move 100 times faster than the tortoise. The tortoise is given 1000 m start. Zeno argued that Achilles could not overtake the tortoise: when Achilles reaches the starting position, T_0, of the tortoise, the tortoise will have moved to T_1, 10 metres away.

When Achilles reaches T_1, the tortoise will have moved to T_2, 0.1 metres away. When Achilles reaches T_2, the tortoise will have moved to T_3, 0.001 metres away. The tortoise will always have moved on!

a **(i)** What is the sum to infinity of the series $1000 + 10 + 0.1 + 0.001 + \cdots$?
(ii) Interpret your answer in the context of the story.

b Assuming Achilles runs at a steady 10 m/s,
(i) write down the start of the series of times it takes him for each *stage* of his run,
(ii) calculate the sum to infinity of these times,
(iii) does this help resolve the paradox?

c Read up on Zeno's other paradoxes.

2 An analogous situation to the above is the problem of finding when the minute hand and hour hand of a clock are coincident.

a Consider 3 o'clock as the starting position.
(i) How much faster is the minute hand than the hour hand?
(ii) How long does it take for the minute hand to reach the starting position of the hour hand?
(iii) How far has the hour hand moved on?
(iv) By considering the sum to infinity, what is the first time after 3 o'clock that the hands of a clock are coincident?

b What is the first time after 6 o'clock that the hands of a clock are coincident?

3 *An unresolved puzzle.* A lamp is switched on. One minute later it is switched off, $\frac{1}{2}$ a minute later it is switched on, $\frac{1}{4}$ a minute later it is switched off, and so on.
Is the lamp on or off after 2 minutes?

Expanding $(1 - x)^{-1}$ and related functions

Reminder

The binomial theorem revisited

Using *sigma notation,* the binomial theorem can be quoted in a very compact form.

$$(x + y)^n = \sum_{r=0}^{n} \binom{n}{r} x^{n-r}y^r$$

Σ stands for *sum* and is pronounced *sigma*. It acts like an instruction set.

1. Create terms using the formula given to the right of Σ by replacing r by each of the integers from 0 to n in turn.
2. Add all these terms together.

So $\quad (x + y)^n = \binom{n}{0}x^{n-0}y^0 + \binom{n}{1}x^{n-1}y + \binom{n}{2}x^{n-2}y^2 + \binom{n}{3}x^{n-3}y^3 + \cdots + \binom{n}{n}x^{n-n}y^n$

where $\quad \binom{n}{r} = \dfrac{n!}{(n-r)!r!}$

Expressions such as $(1 - x)^n$ can be easily expanded using the binomial theorem when $n \geq 0$, $n \in Z$.

$$\begin{aligned}
(1 - x)^5 &= 1 - 5x + 10x^2 - 10x^3 + 5x^4 - x^5 \\
(1 - x)^4 &= 1 - 4x + \ \ 6x^2 - \ \ 4x^3 + \ \ x^4 \\
(1 - x)^3 &= 1 - 3x + \ \ 3x^2 - \ \ \ \ x^3 \\
(1 - x)^2 &= 1 - 2x + \ \ \ \ x^2 \\
(1 - x)^1 &= 1 - x \\
(1 - x)^0 &= 1
\end{aligned}$$

Can we find an expansion for $(1 - x)^{-1}$?

Consider $\dfrac{1}{1 - x}$. If $|x| < 1$ then this can be interpreted as the sum to infinity of the geometric series with a first term of 1 and a common ratio x.

$$(1 - x)^{-1} = \frac{1}{1 - x} = 1 + x + x^2 + x^3 + x^4 + \cdots$$

In sigma notation this would be expressed: $\dfrac{1}{1 - x} = \sum_{r=0}^{\infty} x^r$

It can be proved, but is beyond the scope of this course, that if a series tends to a limit then

- the derivative of the series tends to the derivative of the limit,
- the integral of the series tends to the integral of the limit.

EXERCISE 7A

1 **a** How many terms must be taken before $1 + x + x^2 + x^3 + \cdots + x^{n-1} = (1 - x)^{-1}$ correct to three decimal places, when x equals **(i)** 0.5 **(ii)** 0.1 **(iii)** 0.9?

b By considering the derivative of $1 + x + x^2 + x^3 + \cdots$ find a series which gives the expansion of $(1 - x)^{-2}$, $|x| < 1$. Is it a geometric series?

c Similarly, find a series which gives the expansion of $(1 - x)^{-3}$.

2 Substituting $x = -x$ in $(1 - x)^{-1} = 1 + x + x^2 + x^3 + \cdots$, expand $\dfrac{1}{1 + x}$ in terms of powers of x.
[Note: $|-x| < 1$ is a necessary condition.]

3 **a** By considering the integral of $1 + x + x^2 + x^3 + \cdots$ find a series for $\ln(1 - x)$, $|x| < 1$.
[When $x = 0$ then $\ln(1 - x) = 0$. This fact allows you to evaluate the constant of integration.]
Is it a geometric series?

b Find a series for $\ln(1 + x)$, $|x| < 1$,

c By considering the laws of logs, expand $\ln\left(\dfrac{1 + x}{1 - x}\right)$.

4 **a** By replacing x by x^2 in the expansion for $(1 - x)^{-1}$, find an expansion for $\dfrac{1}{1 - x^2}$.

b Is $\dfrac{1}{1 - x^2}$ an odd or even function?

c Write out the expansion using sigma notation.

5 **a** By replacing x by $1 - x$, in the expansion for $(1 - x)^{-1}$, where $|1 - x| < 1$, expand $\dfrac{1}{x}$ in terms of powers of $(1 - x)$.

b Investigate an expansion for $\ln x$ using integration.
The constant of integration can be computed using the fact that $\ln 1 = 0$.

EXERCISE 7B

1 **a** Expand $\dfrac{1}{1 + x^2}$ using a suitable substitution in the expansion for $\dfrac{1}{1 - x}$.

b Integrate to find an infinite series for $\tan^{-1} x$.
[Remember the constant of integration.]

c The Scottish mathematician James Gregory (1671), by considering the fact that $\tan \dfrac{\pi}{4} = 1$, found an expansion which he could use to evaluate π.

(i) Find Gregory's expansion.

(ii) Use a spreadsheet or graphics calculator to explore the behaviour of the partial sums of the expansion.

2 Expand the following by making suitable substitutions and give the restriction placed on x in each case.

a $(1 - 2x)^{-1}$ **b** $(1 + 3x)^{-1}$ **c** $\left(1 - \dfrac{1}{x}\right)^{-1}$ $\left(= \dfrac{x}{x - 1}\right)$

3 **a** When $-\frac{\pi}{2} < x < \frac{\pi}{2}$, $|\sin x| < 1$. Expand $(1 - \sin x)^{-1}$

b When $-\frac{\pi}{2} < x < \frac{\pi}{2}$, $|\sin^2 x| < 1$. Expand $(\cos x)^{-2}$ [Hint: $\cos^2 x = 1 - \sin^2 x$.]

c Expand $\operatorname{cosec}^2 x$, $0 < x < \frac{\pi}{2}$.

4 **a** Given that $(p - q)^{-1} = p^{-1}\left(1 - \frac{q}{p}\right)^{-1}$ and that $\left|\frac{q}{p}\right| < 1$. Find an expansion for $(p - q)^{-1}$.

b Find an expansion for $(p + q)^{-1}$.

5 Expand the following, stating the conditions under which the expansion is valid.

a $(3 + 4x)^{-1}$ **b** $(2 - 3x)^{-1}$ **c** $(x - 2)^{-1}$

6 **a** Expand $(\sin x - \cos x)^{-1}$. Under what conditions is it valid?

b By considering $(\cos^2 x - \sin^2 x)^{-1}$, expand $(\cos 2x)^{-1}$.

c Investigate $(\cos^2 x + \sin^2 x)^{-1}$.

The above questions touch on a topic referred to as *power series*.
With the aid of a spreadsheet the series can be explored to see
(i) how quickly or slowly the partial sums converge on the *expected* value,
(ii) what happens when we select values outside the valid region.

	A	B	C	D	E
1	0.7	terms	partial sums	target value	difference
2	1	=A1	=B2	=LN(A1)	=C2-D2
3	=A2+1	=-((1-A1)^A3)/A3	=C2+B3	=LN(A1)	=C3-D3

The spreadsheet relates to question 5 above.

7 **a** Expand $\left(1 + \frac{1}{n}\right)^n$ using the binomial theorem when $n = 1, 2, 3, 4$.

b Write down the first five terms of the expansion $\left(1 + \frac{1}{n}\right)^n$.

c Work out the following limits as $n \to \infty$.

(i) $\frac{n(n-1)}{n^2}$ **(ii)** $\frac{n(n-1)(n-2)}{n^3}$

(iii) $\frac{n(n-1)(n-2)(n-3)}{n^4}$ **(iv)** $\frac{n(n-1)(n-2)(n-3)(n-4)}{n^5}$

d Explore the limit of $\frac{n(n-1)(n-2)(n-3)(n-4)\cdots(n-k+1)}{n^k}$ as n tends to infinity.

e Show that $\lim_{n\to\infty}\left(1 + \frac{1}{n}\right)^n = 1 + 1 + \frac{1}{2!} + \frac{1}{3!} + \frac{1}{4!} + \frac{1}{5!} + \frac{1}{6!} + \cdots + \frac{1}{r!}$ + other terms

f Use a calculator to find the partial sum of this series to five decimal places. In 1748 Euler gave this limit as the definition of an important constant. Which constant?

g It can be shown that the sum of the 'other terms' has a limit of zero and that indeed

$$\lim_{n\to\infty}\left(1+\frac{1}{n}\right)^n = e = 1 + 1 + \frac{1}{2!} + \frac{1}{3!} + \frac{1}{4!} + \frac{1}{5!} + \frac{1}{6!} + \dots + \frac{1}{r!} + \cdots$$
$$= \sum_{r=0}^{\infty}\frac{1}{r!}$$

Euler defined the exponential function as

$$e^x = \lim_{n\to\infty}\left(1+\frac{x}{n}\right)^n$$

and the logarithmic function as

$$\ln x = \lim_{n\to\infty} n\left(x^{\frac{1}{n}} - 1\right)$$

Using suitably large values of n and a spreadsheet or calculator, explore the accuracy of the definitions.

h A series is generated so that if $u_n = f(x)$ then $u_{n+1} = \int f(x)\,dx$ ignoring the constant of integration.

(i) Generate such a series when $u_1 = 1$.

(ii) Comment on its partial sums when $x = 1, 2, 3$.

(iii) If the infinite series is the expansion of some function, comment on the derivative of the function.

(iv) Make a conjecture and use a spreadsheet to verify it.

Sigma notation

The basics of sigma notation:

$$\overset{\longleftarrow\; n \text{ terms} \;\longrightarrow}{\sum_{r=1}^{n} f(r) = f(1) + f(2) + f(3) + f(4) + \cdots + f(n)}$$

When $f(r) = 1$, then $f(1) = 1$, $f(2) = 1$, etc.

$$\overset{\longleftarrow\; n \text{ terms} \;\longrightarrow}{\sum_{r=1}^{n} 1 = 1 + 1 + 1 + 1 + \cdots + 1 = n}$$

When $f(r) = r$, then $f(1) = 1$, $f(2) = 2$, etc.

$$\sum_{r=1}^{n} r = 1 + 2 + 3 + 4 + \cdots + n = \frac{n}{2}(n+1)$$

This is the simplest arithmetic series.

Combining sums

When $f(r) = ar + b$

$$\sum_{r=1}^{n}(ar+b) = 1a + b + 2a + b + 3a + b + 4a + b + \cdots + na + b$$
$$= 1a + 2a + 3a + 4a + \cdots + na + b + b + b + b + \cdots + b$$
$$= a(1 + 2 + 3 + 4 + \cdots + n) + b(1 + 1 + 1 + 1 + \cdots + 1)$$
$$= a\sum_{r=1}^{n} r + b\sum_{r=1}^{n} 1$$

$$\sum_{r=1}^{n}(ar+b) = a\sum_{r=1}^{n} r + b\sum_{r=1}^{n} 1 = \frac{an}{2}\left(n+1\right) + bn$$

EXERCISE 8

1 Expand each of the following.

a $\sum_{r=1}^{5} r$ **b** $\sum_{r=1}^{4} r^2$ **c** $\sum_{r=0}^{3} r^3$ **d** $\sum_{r=0}^{4} r!$

e $\sum_{r=1}^{6} (-1)^r$ **f** $\sum_{r=1}^{5} (-1)^r r^2$ [What effect does the factor $(-1)^r$ have on the terms?]

g $\sum_{r=1}^{6} (-1)^{r+1}$ **h** $\sum_{r=1}^{5} (2r+3)$

2 Evaluate each of the following.

a $\sum_{r=1}^{5} (2r+3)$ **b** $\sum_{r=3}^{7} (3r+1)$ **c** $\sum_{r=4}^{8} (3-2r)$ **d** $\sum_{r=-1}^{3} (5-r)$

3 Express each of these series using sigma notation.

a $1 + x + x^2 + x^3 + x^4 + x^5$ **b** $1 - x + x^2 - x^3 + x^4 - x^5$

c $1 + 2x + 4x^2 + 8x^3 + 16x^4$ **d** $1 - 2 + 3 - 4 + 5 - 6$

4 Express each of these series **(i)** in terms of $\sum_{r=1}^{n} r$ and $\sum_{r=1}^{n} 1$, **(ii)** in terms of n.

a $\sum_{r=1}^{n} (3r+2)$ **b** $\sum_{r=1}^{n} (4r+1)$ **c** $\sum_{r=1}^{n} (5r-3)$ **d** $\sum_{r=1}^{n} (4-6r)$

e $\sum_{r=1}^{n} (5-2r)$

5 The arithmetic progression $13 + 19 + 25 + 31 + \cdots + 67$ has an nth term, $u_n = 6n + 7$; 67 is the 10th term and so the progression can be written as $\sum_{r=1}^{10}(6r + 7)$.

Express each of the following arithmetic progressions in sigma notation using a similar method.

a $12 + 16 + 20 + 24 + \cdots + 88$

b $18 + 28 + 38 + 48 + \cdots + 108$

c $25 + 20 + 15 + \cdots + (-50)$

d $19 + 16 + 13 + \cdots + (-11)$

e $1.6 + 1.7 + 1.8 + \cdots + 2.4$

f $\frac{9}{4} + \frac{5}{2} + \frac{11}{4} + \cdots + 22$

6 **a** Check that the progression $-13 + 19 - 25 + 31 - \cdots + 67$ can be written as

$$\sum_{r=1}^{10}(-1)^r(6r + 7)$$

b Use the same strategy to express each of the following in sigma notation.

(i) $-5 + 13 - 21 + 29 - \cdots - 85$

(ii) $-7 + 16 - 25 + 34 - \cdots + 106$

(iii) $-100 + 90 - 80 + \cdots + 10$

(iv) $-26 + 22 - 18 + \cdots - (-46)$

c Verify that $13 - 19 + 25 - 31 + \cdots - 67 = \sum_{r=1}^{10}(-1)^{r+1}(6r + 7)$ and hence express each of the following in sigma notation.

(i) $4 - 5 + 6 - 7 \cdots - 19$

(ii) $8 - 47 + 86 - 125 + \cdots - 437$

(iii) $285 - 268 + 251 - \cdots + 81$

(iv) $61 - 59 + 57 - 55 + \cdots + 29$

7 The geometric progression $3 + 6 + 12 + 24 + \cdots + 1536$ has an nth term, $u_n = 3 \times 2^{n-1}$; 1536 is its 10th term and so the progression can be written as $\sum_{r=1}^{10} 3 \times 2^{r-1}$.

Express each of the following geometric progressions in sigma notation using a similar method.

a $2 + 6 + 18 + 54 + \cdots + 4374$

b $3072 + 1536 + 768 + \cdots + 3$

c $2000 + 200 + 20 + \cdots + 0.002$

d $\frac{9}{4} + \frac{18}{12} + \frac{36}{36} + \cdots + \frac{576}{2916}$

CHAPTER 5 REVIEW

1 Find the first four terms in each of the recurrence relations described.

a $u_{n+1} = 3u_n + 2; \quad u_0 = 3$

b $u_{n+1} = ku_n + p; \quad u_0 = 32$

2 Find the fixed point in each of these recurrence relations and say whether it is stable or not.

a $u_{n+1} = 5u_n + 1$

b $u_{n+1} = 0.5u_n + 6$

3 An arithmetic sequence starts 3, 7, 11, …

a Find an expression for the nth term of the sequence.

b Is 91 a term in the sequence?

4 The fourth term of an arithmetic sequence is −3 and the tenth term is −15.

a Identify the sequence.

b For what value of n is $u_n = -35$?

5 An arithmetic sequence has a common difference of 7 and the 25th term equals 300. What is the first term?

6 Find the sum to the first 30 terms of the arithmetic sequence which starts 2, 2.5, 3, 3.5, …

7 The sum of the first 10 terms of an arithmetic sequence is 155. The sum to 11 terms is 187. Identify the sequence.

8 How many terms must be added before the sum of the sequence 4, 4.25, 4.5, … exceeds 100?

9 a Find an expression for the nth term of the geometric sequence 4, 12, 36, …

b Which term in this sequence is closest to 1000?

10 The first term of a geometric sequence is 17. The ninth term is 111 537. Find the common ratio.

11 The common ratio of a geometric sequence is 0.6. The ninth term is 6561. Find the first term.

12 Find the sum to eight terms of the geometric sequence:

a 3, 15, 75, …

b 400, 100, 25,…

13 The first two terms of a geometric series add up to 36. The fourth and fifth terms add up to 4.5. Find the sum of the first 10 terms.

14 a When will the sum of the series 20 + 10 + 5 … first exceed 39.9?

b State why you know the sum to infinity of this series exists, and find it.

15 a Write down an expansion for $\frac{1}{1-x}$ given that $|x| < 1$.

b Hence find an expansion for $\frac{1}{2-3x}$ given that $|x| < \frac{2}{3}$.

16 a State the limit of $\left(1+\frac{1}{n}\right)^n$ as n tends to infinity.

b Hence find the limit of $\left(1+\frac{3}{n}\right)^n$ as n tends to infinity.

17 Find an expression for $3\sum_{r=1}^{n} r + \sum_{r=1}^{n} 2$ in terms of n.

CHAPTER 5 SUMMARY

1 (i) A *sequence* is an ordered list of numbers where each term is obtained according to a fixed rule.

(ii) A *series*, or *progression*, is a sum, the terms of which form a sequence.

(iii) The nth term of a sequence is denoted by u_n.

(iv) A sequence can be defined by a recurrence relation where u_{n+1} is given as a function of lower/earlier terms.

(v) A first-order recurrence relation is where u_{n+1} is given as a function of u_n.

(vi) A first-order linear recurrence relation is where $u_{n+1} = ru_n + d$, where r and d are constants.

(vii) A sequence can be defined by a formula for u_n given in the form of a function, $u_n = f(n)$.

2 (i) When $r = 1$ and $d \neq 0$, i.e. $u_{n+1} = u_n + d$, then the sequence is called *arithmetic*.

(ii) For an arithmetic sequence $u_{n+1} - u_n = u_{n+2} - u_{n+1} = d$ and u_1 is denoted by a.

(iii) The nth term of an arithmetic sequence is given by $u_n = a + (n-1)d$.

(iv) The sum to n terms of an arithmetic sequence is given by $S_n = \frac{n}{2}\left(2a + (n-1)d\right)$.

3 (i) When $r \neq 0, 1$ and $d = 0$, i.e. $u_{n+1} = ru_n$, then the sequence is called *geometric*.

(ii) For a geometric sequence $u_{n+1} \div u_n = u_{n+2} \div u_{n+1} = r$ and u_1 is denoted by a.

(iii) The nth term of a geometric sequence is given by $u_n = ar^{n-1}$.

(iv) The sum to n terms of a geometric sequence is given by $S_n = \frac{a(1-r^n)}{1-r}$.

(v) When $|r| < 1$ then $r^n \to 0$ as $n \to \infty$ and the *sum to infinity*, which is the limit to which the partial sums tend, is denoted by S_∞ and $S_\infty = \frac{a}{1-r}$

4 When $|x| < 1$ then $\frac{1}{1-x} = 1 + x + x^2 + x^3 + x^4 + x^5 + x^6 \cdots$

5 (i) The sequence defined by $u_n = \left(1 + \frac{1}{n}\right)^n$ begins 2, 2.25, 2.370 37, 2.441 41 … (5d.p.)

(ii) As $n \to \infty$, $u_n \to 1 + \frac{1}{1!} + \frac{1}{2!} + \frac{1}{3!} + \cdots = e = 2.718\,281\,828\ldots$

6 (i) $\sum_{r=1}^{n} f(r) = f(1) + f(2) + f(3) + \cdots + f(n)$

(ii) $\sum_{r=1}^{n} 1 = 1 + 1 + 1 + \cdots + 1 = n$

(iii) $\sum_{r=1}^{n} r = 1 + 2 + 3 + \cdots + n = \frac{n}{2}\left(n + 1\right)$

(iv) $\sum_{r=1}^{n} ar + b = a\sum_{r=1}^{n} r + b\sum_{r=1}^{n} 1$

Answers

CHAPTER 1

Exercise 1A (page 3)

1 **a** $3x + 5x \neq 7x$, T
b $3 + 6 \leq 7$, F
c $4^3 < 3^4$, T
d $\sin x < -1$, $\sin x > 1$, F
e $\frac{d}{dx}e^x \neq e^x$, F
f $2x + 1$ is not odd given $x \in W$, F

2 **a** $\exists$a whole number s.t. $2x + 1 < 5$
b $\exists x$ s.t. $\frac{1}{x}$ does not exist
c $\exists x$ s.t. $x^2 \leq 0$
d $\exists n$ s.t. $2^n + 1$ is not prime

3 **a** All whole numbers of the form $2x + 1$ are odd, T
b $\forall x$, $2^x \leq 3^x$, F
c All natural numbers ≥ 1, T
d All prime numbers are odd, F
e $\forall x(> 0)\ \ln x < \ln(x + 1)$, T
f $\forall x\ \sqrt{x} \in R$, F

4 **a** In: if *a shape is not an isosceles triangle* then *it will not have two equal sides*
Cn: if *a shape has two equal sides* then *it is an isosceles triangle*
Cp: if *a shape does not have two equal sides*, then *it is not an isosceles triangle.*
b In: if *a number is not divisible by 2* then *it is not even*
Cn: if *a number is even*, then *it is divisible by 2*
Cp: if *a number is not even*, then *it is not divisible by 2*
c In: if $2x + 4 \ngtr 12$ then $x \ngtr 4$
Cn: if $x > 4$ then $2x + 4 > 12$
Cp: If $x \ngtr 4$ then $2x + 4 \ngtr 12$
d In: if $\sqrt{x} \ngtr 1$ then $x \ngtr 1$
Cn: if $x > 1$ then $\sqrt{x} > 1$
Cp: if $x \ngtr 1$ then $\sqrt{x} \ngtr 1$.

5 **a** T, T, two-way **b** T, T, two-way
c T, T, two-way **d** T, T, two-way
e T, F **f** T, T, two-way
g T, F **h** T, T, two-way
i T, F

6 **a, d, f, h**

7 **a** $x = 0$ **b** $x = 1$ **c** $x = 0$
d $x < 0$ **e** $x = \frac{25}{3}$
f any positive a and negative b
g $x = 0$

8 **a** $x = 0$ **b** $x = 1$ **c** any a; $b = 0$
d $A = B = 0$ **e** $x = 0$ **f** any $a = b$
g any $x \geq \frac{\pi}{2}$

Exercise 1B (page 5)

1 **a** dividing by $(x - y)$, which is zero
b dividing by $\ln\frac{1}{2}$ without reversing the inequality sign (since $\ln\frac{1}{2}$ is negative).

2 **a** L3: $x = \pm 4$
b L1: not a property of parallelograms
c L5: triangle is right angled at P
d L2: $-9x < -20$
e L2: fraction upside down;
L4: false, c/ex: $\frac{\sqrt{8}}{\sqrt{2}} = 2$

3 **a** 2, 4 **b** 2, 5
c 4, 6 **d** 2, 3, 4, 5

4 **a** equivalence
b c/ex (2nd): $x = -7$
c c/ex (2nd): ABC could be a triangle
d c/ex (1st): $a = 2 + \sqrt{2}$, $b = 2 - \sqrt{2}$
e c/ex (1st): $a = 2 + \sqrt{2}$, $b = 2 - \sqrt{2}$
f equivalence

5 **a** c/ex: $u_4 = 13$
b c/ex: rectangle
c c/ex: 1 June is Wednesday
d c/ex: $5|(3 + 2)$, $5|(2 + 8)$, but 5 does not divide $3 + 8$
e c/ex: $\int x^{-1}\,dx \neq \frac{1}{0}x^0 + c$
f c/ex: $S(6) = 1 + 2 + 3 = 6$, i.e. $S(6) = 6$
g c/ex: $6|3 \times 4$, but $6 \nmid 3$, $6 \nmid 4$

6

k	2	3	5	7	11
$2^k - 1$	3	7	31	127	$2047 = 23 \times 89$

Exercise 2A (page 10)

1 **a** hint: reverse and add
b hint: reverse and add

2 **a** hint: multiply by 3 and subtract
b hint: multiply by 3 and subtract

3 **a** $2k$ or $2k + 1$, $k \in W$
b hint: consider two cases, i.e. where k is even/odd
c **(i)** $3k$, $3k - 1$, $3k + 2$, $k \in N$
(ii) hint: consider three cases
d proof

4 hint: factorise and examine the factors

5 hint: consider the three *sloping* triangles, use Pythagoras and eliminate the *sloping* variables

6 proof
7 start: n is odd $\Rightarrow n = 2k + 1, n \in W$
8 proof

Exercise 2B (page 11)

1 hint: what does 6×6 have as a units digit?
2 hints: $3^{2n} - 1 = (3^n - 1)(3^n + 1)$
$x^n - 1 = (x - 1)(x^{n-1} + {}^{n-1}C_1 x^{n-2}(-1)^1 + \ldots + (-1^{n-1}))$
$3^{2n} + 7 = 3^{2n} - 1 + 8$
3 hint: $1099 \ldots 9989 = 1100 \ldots 0000 - 11$
4 hint given
5 hints: $125 = 7 \times 17 + 6$ but it will be easier if you write this as $125 = 7 \times 18 + (-1) = 7k - 1$, i.e. treat the remainder as -1 rather than 6. Then $5^{99} = (5^3)^{33} = (7k - 1)^{33} = \ldots$
6 hint: factorise
7 **b** product of three consecutive numbers plus the middle number is equal to the cube of the middle number
8 hint given
9 hint: consider k even and k odd separately
10 hint: $\cos\left(x + \frac{\pi}{3}\right) = \cos x \cos\frac{\pi}{3} - \sin x \sin\frac{\pi}{3}$
$= \frac{1}{2}\cos x - \frac{\sqrt{3}}{2}\sin x$

Exercise 3A (page 14)

1 assume $\sqrt{3} = \frac{a}{b}$, $a, b \in N$
2 assume $\sqrt{7} = \frac{a}{b}$, $a, b \in N$
3 assume $\frac{1}{2}(6\sqrt{3} - 1) = \frac{a}{b}$, $a, b \in N$ and make $\sqrt{3}$ the subject
4 consider a factor of 2
5 consider a factor of 7
6 hint given
7 prove that if $x = 75$ then x is not even
8 **a** assume $\exists x$ where x^2 is even and x is odd
b examine x^2 is not odd and x is not odd and assume what you've proved in **8a**
9 assume the negation: $\exists n$ for which $n^2 + n$ is odd
10 assume $a \neq 3$ and $b \neq 4$; this leads to the statement that $\sqrt{2}$ is rational
11 assume both numbers are rational, leading to rational sum
12 assume both A and B are odd while AB is even
13 use contrapositive: 'If two lines are not parallel then they cannot both be perpendicular to a third line in the plane.' Draw a diagram with two lines intersecting at an angle.

Exercise 3B (page 17)

1 use example for $\sqrt{2}$ as a model
2 **b** c/ex: $\sqrt{8} + \sqrt{2}$
3 hint given
4 **b** example: $p = 17$
5 **a** $p = 37$ **b** 40 and 112
6 5
7 **a** 73 **b** 9 and 104

Exercise 4 (page 20)

1 goal: $u_{k+1} = (k + 1)^2 + (k + 1) - 1 = k^2 + 3k + 1$
2 goal: $u_{k+1} = 2(k + 1)(k + 2) = 2k^2 + 6k + 4$
3 goal: $u_{k+1} = 3 \times 2^k - 1$
4 **a** conjecture:
$S_n = 1 + 3 + 5 + \cdots + (2n - 1) = n^2$
b goal: $S_{k+1} = (k + 1)^2$
5 **a** $u_n = u_{n-1} + 3(n - 1) + 1 = u_{n-1} + 3n - 2$
so $u_{n+1} = u_n + 3n + 1$
b goal: $u_{k+1} = \frac{1}{2}(k + 1)(3k + 2)$
6 **a** goal: $2^{k+1} > k + 1$; use
$2k = k + k > k + 1$ for $k > 1$
b goal: $3^{k+1} > 2^{k+1}$
c goal: $2^{k+1} > (k + 1)^2$
7 **a** goal: $u_{k+1} = (k + 1)^2 - 4 = k^2 + 2k - 3$
b hint: reverse and add
$2S = (2n + 4) + (2n + 4) + \cdots + (2n + 4)$ $k - 2$ times
8 assume $2^{3k-1} = 7m$; goal $2^{3k+3} - 1 = 7n$
9 **a** assume $3^{2k} - 5 = 4m$; goal $3^{2k+2} - 5 = 4n$
b direct: factorise $x^n - 1 = (x - 1)(\ldots)$ and then set $x = 9$
c $3^{2n} - 5 = 3^{2n} - 1 - 4 = 9^n - 1 - 4 \ldots$
10 goals:
a $\sum_{r=1}^{k+1} F_r = F_{k+3} - 1$ **b** $\sum_{r=1}^{k+1} F_{2r} = F_{2k+3} - 1$
c $\sum_{r=1}^{k+1} F_{2r-1} = F_{2k+2} - 1$ **d** $\sum_{r=1}^{k+1} F_r^2 = F_{k+1}F_{k+2}$
e $F_{k+1}^2 = F_k \times F_{k+2} - 1$
f $F_{k+1}^2 + F_{k+2}^2 = F_{2k+3}$
11 **a** hint: $\frac{d}{dx}(x^{n+1}) = \frac{d}{dx}(x.x^n)$
b The induction process depends on establishing the truth of a smallest value of n, clearly impossible for $n \in Z$ or Q.
12 $n \geq 4$, goal: $(k + 1)! > 2^{k+1}$,
i.e. $(k + 1) \times k! > 2^{k+1}$
13 assume $k = 4m_1 + 7m_2$;
goal: $k + 1 = 4m_1' + 7m_2'$;
separate into k even and k odd
14 hint given

Exercise 5 (page 24)

1 hint: $9^n = (8 + 1)^n$ and expand

2 a hint: multiply out the brackets

b hint: $p + q + r \geq 3\sqrt[3]{pqr}$

c hint: $pq + qr + rp \geq 3\sqrt[3]{p^2q^2r^2}$

d hint; combine and let $pqr = t^3$; then solve for t

3 hint: let the three numbers be $\frac{a}{2}, \frac{a}{2}$ and b.

4, 5 hints given

Review (page 25)

1 hint: reverse and add

2 hint: expand brackets

3 kite

4 assume $\sqrt{11} = \frac{a}{b}$ where $a, b \in Z$

5 goal: $S_{k+1} = \frac{1}{2}(k + 1)(3k + 4)$

6 a assume $k(k^2 - 1)(3k + 2) = 24m$, testing $(k + 1)((k + 1)^2 - 1)(3(k + 1) + 2)$ so expand each of these expressions

b hint: consider separately n even and odd.

CHAPTER 2.1

Exercise 1A (page 29)

1 a $\frac{1}{5}x^{\frac{-4}{5}}$ b $\frac{4}{3}x^{\frac{1}{3}}$

c $-\frac{1}{x\sqrt{2x}}$ d $\frac{1}{2\sqrt{x-1}}$

2 a (i) $(x + 2)^2 + 1$ (ii) $\sqrt{x - 1} - 2$

(iii) $2x + 4, \frac{1}{2\sqrt{x-1}}$

b (i) $(x + 3)^2 - 10$ (ii) $\sqrt{x + 10} - 3$

(iii) $2x + 6, \frac{1}{2\sqrt{x+10}}$

c (i) $2\left(x - \frac{1}{2}\right)^2 + \frac{1}{2}$ (ii) $\frac{1}{2} + \frac{1}{2}\sqrt{2x - 1}$

(iii) $4x - 2, \frac{1}{2\sqrt{2x-1}}$

d (i) $3\left(x + \frac{1}{3}\right)^2 - \frac{1}{3}$ (ii) $\frac{1}{3}\sqrt{3x + 1} - \frac{1}{3}$

(iii) $6x + 2, \frac{1}{2\sqrt{3x+1}}$

3 a inverse function method

b $\frac{1}{2x}$

4 a $\frac{-1}{\sqrt{1-x^2}}$ b $\frac{1}{1+x^2}$

Exercise 1B (page 30)

1 a $\frac{1}{1+x^2}$ b $\frac{3}{\sqrt{(-3x)(3x+2)}}$

2 a (i) $(x + 3)^2 - 8$ (ii) $x = \sqrt{y + 8} - 3$

(iii) $\frac{1}{2\sqrt{y+8}}$ (iv) $2x + 6$

b (i) $(x + 1)^2 - 6$ (ii) $x = \sqrt{y + 6} - 1$

(iii) $\frac{1}{2\sqrt{y+6}}$ (iv) $2x + 2$

c (i) $3\left(x - \frac{2}{3}\right)^2 - \frac{13}{3}$ (ii) $x = \frac{2}{3} + \frac{1}{3}\sqrt{3y + 13}$

(iii) $\frac{1}{2\sqrt{3y+13}}$ (iv) $6x - 4$

d (i) $2(x + 1)^2 - 2$ (ii) $x = \sqrt{\frac{y+2}{2}} - 1$

(iii) $\frac{\sqrt{2}}{4\sqrt{y+2}}$ (iv) $4x + 4$

Exercise 2 (page 32)

1 a $\frac{2x}{\sqrt{1-x^4}}$ b $\frac{1}{x^2+4x+5}$

c $\frac{-1}{x\sqrt{x^2-1}}$ d $\frac{-1}{2\sqrt{x}(x+1)}$

e $\frac{1}{x\sqrt{x^2-1}}$ f $\frac{-a}{\sqrt{1-a^2x^2}}$

2 a $\frac{e^x}{\sqrt{1-e^{2x}}}$ b $\frac{-2(x+2)}{\sqrt{1-(x+2)^4}}$

c $\frac{-1}{\sqrt{1-x^2}}$ d $\frac{1}{\cos x\sqrt{\cos 2x}}$

e $\frac{1}{\sqrt{a^2-x^2}}$

3 a $\frac{-2e^{2x}}{\sqrt{1-e^{4x}}}$ b -1

c $\frac{1}{x^2+2x+2}$ d $\frac{-1}{x\sqrt{1-(\ln 3x)^2}}$

e $\frac{1}{x\sqrt{9x^2-1}}$

4 a $\frac{1}{2\sqrt{x}(1+x)\tan^{-1}\sqrt{x}}$

b $\frac{1}{2x\sqrt{x-1}\sin^{-1}\left(\frac{-1}{\sqrt{x}}\right)}$

c $\frac{e^x}{\sin^{-1}(e^x)\sqrt{1-e^{2x}}}$ d $\frac{e^{\sin^{-1}x}}{\sqrt{1-x^2}}$

5 a $-\frac{1}{2}e^{\frac{\pi}{4}}$ b $-\frac{2}{\pi}$

c $\frac{64}{125}$ d $\frac{1}{8}$

Answers

6 a $\frac{8\sqrt{3}}{\pi}$ b $\frac{9(4+\pi)^2}{32}$

c $\frac{2\sqrt{3}}{\pi}$ d $e^{\frac{\pi}{4}}$

7 a $x > 0$

9 a $-\frac{1}{2} \le x \le \frac{1}{2}$

Exercise 3A (page 33)

1 a $\frac{x^2}{\sqrt{1-x^2}} + 2x\sin^{-1}x$

b $\frac{2x^2}{\sqrt{1-x^4}} + \sin^{-1}x^2$

c $\frac{1}{2\sqrt{x}}\cos^{-1}x - \sqrt{\frac{x}{1-x^2}}$

d $\frac{1}{2\sqrt{1-x}} + \frac{\sin^{-1}\sqrt{x}}{2\sqrt{x}}$

2 a $1 + 2x\tan^{-1}x$

b $e^x\left[\sin^{-1}x + \frac{1}{\sqrt{1-x^2}}\right]$

c $e^{2x}\left[2\cos^{-1}\left(\frac{x}{2}\right) - \frac{1}{\sqrt{4-x^2}}\right]$

d $\frac{\ln x}{x^2+1} + \frac{\tan^{-1}x}{x}$

3 a $\frac{x-(1+x^2)\tan^{-1}x}{x^2(1+x^2)}$

b $\frac{2x-\sqrt{1-x^2}\sin^{-1}x}{2x\sqrt{x(1-x^2)}}$

c $\frac{4x+3\sqrt{1-4x^2}\cos^{-1}2x}{-2x^{\frac{5}{2}}\sqrt{1-4x^2}}$

d $\frac{x-2(x^2+2x+2)\tan^{-1}(x+1)}{x^3(x^2+2x+2)}$

4 a $\frac{\sqrt{1-x^2}\sin^{-1}x - x}{\sqrt{1-x^2}(\sin^{-1}x)^2}$

b $\frac{2x\sqrt{2x-x^2}\cos^{-1}(x-1)+x^2}{\sqrt{2x-x^2}(\cos^{-1}(x-1))^2}$

c $\frac{e^x\sqrt{1-4x^2}\sin^{-1}2x - 2e^x}{\sqrt{1-4x^2}(\sin^{-1}2x)^2}$

d $\frac{(1+x^2)\tan^{-1}x - x\ln x}{x(1+x^2)(\tan^{-1}x)^2}$

5 a $-\frac{4}{5}$ b $\frac{6-8\sqrt{3}}{13}$

6 a $(1, \tan^{-1}e)$ b $\left(e, \cos^{-1}\left(\frac{1}{e}\right)\right)$

Exercise 3B (page 34)

2 $\frac{6(1+\cos x)}{5+4\cos x}$ 4 0

Exercise 4A (page 36)

1 a $-\frac{x+2y}{2x+y}$ b $2xy$

c $\frac{5-4x}{4(y+1)}$ d e^{-y}

e $\frac{y-2x}{6y-x}$ f $\frac{e^x-\tan y}{x\sec^2 y}$

g $-\left(\frac{y}{x}\right)^{\frac{3}{5}}$ h $\frac{x+y-x^2-1}{x^2+1}$

i $\frac{5x-2y}{2x-3y}$

j $\sqrt{\frac{1-y^2}{1-x^2}} - 6x^2\sqrt{1-y^2}$

2 a (i) $\frac{1}{\cos y}$ (ii) $\frac{1}{\sqrt{1-x^2}}$

b (i) $-\frac{1}{\sin y} = -\frac{1}{\sqrt{1-x^2}}$

(ii) $\frac{1}{\sec^2 y} = \frac{1}{1+x^2}$

3 $\frac{2x^3+y}{x(1-2xy)}$

5 $10x + 11y = 32$

9 $8x + 21y = 21$

Exercise 4B (page 37)

1 $-\frac{2}{3}$

3 a $(2, 2), (2, -2)$ b $-\frac{1}{2}, -1; 71.6°$

5 a $(20, 6), (20, -6), (-20, 6), (-20, -6)$

6 $\ln(x-y)$

Exercise 5 (page 38)

1 a $-\frac{2x+y}{x}, \frac{2(x+y)}{x^2}$

b $-\frac{3x^2}{2y}, \frac{-3x(4y^2+3x^3)}{4y^3}$

c $-\frac{1}{6\sqrt{xy^2}}, \frac{3y^3-2\sqrt{x}}{36x\sqrt{xy^5}}$

d $\frac{y}{2y-x}, \frac{2y(y-x)}{(2y-x)^3}$

e $\frac{e^x-y+1}{x+1}, \frac{2(y-1)+e^x(x-1)}{(x+1)^2}$

f $\frac{y^2}{1-2xy}, \frac{2y^3(2-3xy)}{(1-2xy)^3}$

g $\frac{x+y-1}{x+y+1}, \frac{4(x+y)}{(x+y+1)^3}$

h $\dfrac{2x}{1+\ln y}, \dfrac{2y(1+\ln y)^2-4x^2}{y(1+\ln y)^3}$

i $\dfrac{2(x+y)}{e^y-2(x+y)}, \dfrac{2e^y(e^y-2(x+y)^2)}{(e^y-2(x+y))^3}$

j $\dfrac{x\cos x+\sin x+y}{1-x}$,

$\dfrac{2y+2\sin x+2\cos x-x\sin x+x^2\sin x}{(1-x)^2}$

k $\dfrac{\cos(x+y)}{1-\cos(x+y)}, \dfrac{\sin(x+y)}{(\cos(x+y)-1)^3}$

2 $-e(e+1)$, concave up

3 -1, concave up

6 $-2, 18$

7 $2e^{\frac{\pi}{3}}(1+\sqrt{3}), 4e^{\frac{\pi}{3}}(4+\sqrt{3})$

8 $(1, 2), (-1, -2); (2, 1)(-2, -1)$

9 $(1, 4)$ maximum turning point; $(-1, -4)$ minimum turning point

Exercise 6 (page 40)

1 a $(2\ln 5)5^{2x}$

b $(x+1)^{x-1}\left(\left(\dfrac{x-1}{x+1}\right)+\ln(x+1)\right)$

c $\sin 2xe^{\sin^2 x}$

d $\ln 3e^x 3^{e^x}$

e $(\ln\cos x - x\tan x)(\cos x)^x$

2 a $(x+2x\ln x)x^{x^2}$

b $3x^2\ln\pi.\pi^{x^3}$

c $\dfrac{e^x((x-1)\sin x+x\cos x)}{x^2}$

d $(\cos x - x\sin x - x\cos x)e^{-x}$

e $(1-x^3)^{\sin x}\left(\cos x\ln(1-x^3)-\dfrac{3x^2\sin x}{1-x^3}\right)$

7 $y=9x-9$

8 $-\dfrac{11}{288}$

9 $\dfrac{9}{10}$

10 a $\dfrac{7x^2+18x-1}{12(x+1)^{\frac{1}{2}}(x-1)^{\frac{2}{3}}(x+2)^{\frac{5}{4}}}$

b $\dfrac{(5x^2-24x+9)(2x+3)^{\frac{3}{2}}}{3x^2(x-1)^{\frac{5}{3}}}$

Exercise 7A (page 42)

1 a $2x+y=3$

b $xy=9$

c $y^2=20x$

d $\dfrac{x^2}{16}+\dfrac{y^2}{9}=1$

e $\dfrac{x^2}{25}-\dfrac{y^2}{144}=1$

f $\dfrac{(x-3)^2}{4}+\dfrac{(y-2)^2}{9}=1$

g $x^2+y^2=2$

h $16x+15y=17$

Exercise 8A (page 44)

1 a $-\dfrac{1}{t^2}, \dfrac{2}{t^3}$

b $\dfrac{1}{2t^2}, -\dfrac{1}{2t^4}$

c $\dfrac{1+\sin t}{1+\cos t}, \dfrac{(\cos t+\sin t+1)}{(1+\cos t)^3}$

d $\dfrac{8t}{9t^2-1}, \dfrac{-8-72t^2}{(9t^2-1)^3}$

e $\dfrac{\sin\theta}{\cos\theta-1}, \dfrac{1}{(1-\cos\theta)^2}$

2 a $(-1, 3)$

b minimum turning point

3 a $\dfrac{t^4-1}{t^4+1}, \dfrac{4t^6}{(t^4+1)^3}$

b $(0, 2)$ minimum turning point

5 max at $\left(\dfrac{e^{\frac{\pi}{4}}}{\sqrt{2}}, \dfrac{e^{\frac{\pi}{4}}}{\sqrt{2}}\right)$; min at $\left(-\dfrac{e^{\frac{5\pi}{4}}}{\sqrt{2}}, -\dfrac{e^{\frac{5\pi}{4}}}{\sqrt{2}}\right)$

6 b $(1, 0), (-3, 0), \left(\dfrac{3}{2}, \dfrac{\sqrt{3}}{2}\right), \left(\dfrac{3}{2}, -\dfrac{\sqrt{3}}{2}\right)$

Exercise 8B (page 45)

1 a $(0, \sqrt{2})$, end point minimum

b $x=\sqrt{2}$

3 b $\left(0, \pm\dfrac{1}{\sqrt{2}}\right)\left(\pm\dfrac{1}{\sqrt{2}}, 0\right)$

4 a $2\cos\theta, -2\tan\theta$

5 a $(0, 0)$ min, $\left(2^{\frac{1}{3}}, 2^{\frac{2}{3}}\right)$ max

b $\dfrac{d^2y}{dx^2}=0$

Review (page 46)

1 a $\dfrac{3x^2}{2}, \sqrt[3]{2x+5}$

b $\dfrac{2}{3(2x+5)^{\frac{2}{3}}}$

2 $\dfrac{\sqrt{1-x^2}}{4-x}$

3 $\dfrac{5}{\sqrt{1-25x^2}}$

4 $\dfrac{9}{4\pi}$

5 $\frac{1}{2}e^{\frac{x}{2}}\left(\cos^{-1}2x-\dfrac{4}{\sqrt{1-4x^2}}\right)$

6 $-\frac{1}{2}$

7 $-\dfrac{y^2}{x^2}; \dfrac{2y^3+2xy}{x^4}$

8 $\ln 2+\frac{1}{4}$

9 $y=-\dfrac{x^2-4x+5}{x^2-4x+3}$

10 a (i) $y=x+4, 3y=2x-1$ (ii) $(-13, -9)$

b $\dfrac{d^2y}{dx^2}=0 \Rightarrow t=0 \Rightarrow x, y\to\infty$

CHAPTER 2.2

Exercise 1 (page 50)

(unless otherwise stated, units used are m s^{-1} and m s^{-2})

1 141.9, 62.4°

2 **a** (28, 24), (81, 45)
b (i) $18\sqrt{5}$, 26.6° (ii) $24\sqrt{10}$, 18.4°
c (i) $6\sqrt{26}$, 11.3° (ii) $30\sqrt{2}$, 8.1°

3 **a** (i) $(3 + \ln 2, 4 + \ln 3)$ (ii) $(3 + \ln 3, 4 + \ln 4)$
b (i) $\frac{\sqrt{13}}{6}$, 33.7° (ii) $\frac{5}{12}$, 36.9°
c (i) $\frac{\sqrt{97}}{36}$, 204.0° (ii) $\frac{\sqrt{337}}{144}$, 209.4°

4 **a** 56.1; 86.6° **b** 7.6; 89.2°

5 **a** $t = 1$
b (i) (−3, −2) (ii) $6\sqrt{2}$; 45°

6 **a** $t = 2$
b (−8, −20); $2\sqrt{61}$; 39.8°

7 **a** (i) $\sqrt{2\left(e^{2t} + \frac{1}{e^{2t}}\right)}$ (ii) $\frac{e^{2t} + 1}{e^{2t} - 1}$
b $v \to \infty$, $\alpha \to 45°$

8 **a** 135° **b** $\frac{\sqrt{2}}{8}$, $\frac{3\sqrt{2}}{256}$

Exercise 2A (page 53)

1 **a** $0.5L\,\text{m}^3$ **b** $\frac{1}{2}$ **c** $1 - \frac{t}{7200}$
d $2 - \frac{t}{3600}$
e (i) $\frac{14\,399}{14\,400}$ m^3/h (ii) $\frac{14\,399}{7200}$ m/h

2 **a** $\frac{dy}{dt} = 4\frac{dy}{dx}$ **b** (i) 108 (ii) 12

3 $4(4x - 3)^5$

4 **a** 1 **b** (i) $A = \pi r^2$ (ii) $2\pi r$
c 8π cm^2/h

5 **a** (i) $V = x^3$ (ii) $3x^2$
b 0.6 cm^3/min

6 **a** $4\pi r^2$ **b** 0.3 cm/s

Exercise 2B (page 54)

1 **a** $t + 1$ **b** 15 min
c (i) 60 (ii) 80

2 $-3\,\text{N m}^{-2}\,\text{s}^{-1}$

3 **a** (i) $h = 4r$ (ii) $V = \frac{4}{3}\pi r^3$ (iii) $\frac{dV}{dr} = 4\pi r^2$
b $\frac{4}{\pi}$ cm/min **c** $\frac{25}{\pi}$ cm/min

4 200 m^3/min

5 $\frac{2}{3\pi}$ m/s

6 **a** $\frac{3}{40}$ cm/s **b** $\frac{1}{10}$ cm/s **c** $\frac{3\pi}{2}$ cm^2/s

8 $-15\sqrt{3}$ cm/s

9 3.6 units/s

10 1.6 cm/s

11 60 m/s

Review (page 57)

1 **a** $48\sqrt{5}$ **b** 63.4°
c (i) $48\sqrt{10}$ **d** 71.6°

2 **a** 8 units/min **b** 8 u^2/min

CHAPTER 3

Exercise 1A (page 61)

1 **a** $\sin^{-1}\frac{x}{5} + c$ **b** $\frac{1}{5}\tan^{-1}\frac{x}{5} + c$
c $\sin^{-1} x + c$ **d** $\frac{1}{\sqrt{3}}\tan^{-1}\frac{x}{\sqrt{3}} + c$
e $\frac{1}{3}\sin^{-1}\frac{x}{2} + c$ **f** $\frac{1}{12}\tan^{-1}\frac{x}{4} + c$
g $\frac{1}{15}\tan^{-1}\frac{x}{3} + c$ **h** $\frac{1}{4}\sin^{-1}\frac{x}{3} + c$

2 **a** $\frac{\pi}{2}$ **b** $\frac{\pi}{3}$ **c** $\frac{\pi}{6\sqrt{3}}$
d $\frac{\pi}{12}$ **e** $\frac{\pi}{48}$ **f** $\frac{1}{2}\sin^{-1}\frac{2}{3}$
g $\frac{\pi}{3\sqrt{3}}$ **h** $\frac{\pi}{18\sqrt{3}}$

3 $a = 8$

4 proof

5 $a = \frac{1}{\sqrt{2}}$

Exercise 1B (page 62)

1 **a** $\frac{\pi}{18}$ **b** $\frac{1}{4}\sin^{-1}\frac{4}{5}$ **c** $\frac{\pi}{6\sqrt{6}}$
d $\frac{2\pi}{9\sqrt{3}}$ **e** $\frac{\pi}{3\sqrt{3}}$ **f** $\frac{5}{12}\tan^{-1}\left(\frac{3}{2}\right)$
g $\frac{\pi}{4\sqrt{2}}$ **h** $\frac{7\sqrt{2}}{2}\ln(\sqrt{2} + \sqrt{3})$

2 **a** $-\sin^{-1}\frac{x}{a} + C_1$ **b** $\cos^{-1}\frac{x}{a} + C_2$
c $C = \frac{\pi}{2}$

3 **a** proof **b** $\ln|x + \sqrt{x^2 + k}| + c$
c (i) $\ln\left(\frac{2 + \sqrt{7}}{\sqrt{3}}\right)$ (ii) $2\ln(\sqrt{2} + 1)$
(iii) $\frac{1}{2}\ln\left(\frac{4 + \sqrt{11}}{5}\right)$ (iv) $\frac{1}{2}\ln\left(\frac{1 + \sqrt{5}}{2}\right)$
(v)/(vi) not defined for the given limits

Exercise 2 (page 64)

1 $\ln\left|\frac{(x-2)^2}{x-1}\right| + c$

2 $\ln|x-2| - \frac{2}{x-2} + C$

3 $\ln\frac{32}{27}$

4 $\frac{14}{15}$

5 a $\ln\left|\frac{(x+2)^3}{(2x-3)}\right| + C$ b $-\frac{1}{x-3} + c$

c $\ln\left|\frac{(x-1)(x-3)}{(x-2)^2}\right| + c$ d $2\ln\left|\frac{x-2}{x+3}\right| + c$

e $\frac{1}{2}\ln\left|\frac{(x-2)^2}{x^2-1}\right| + c$

f $\ln|x+1| + \frac{1}{x+1} + c$

6 a $\ln\left(\frac{49}{32}\right)$ b $\ln\left(\frac{256}{135}\right)$ c $\frac{1}{20}$

d $\frac{2}{5}$ e $3\ln 2 + \frac{1}{2}$ f $\ln\left(\frac{27}{5}\right)$

Exercise 3A (page 66)

1 $\ln\left|\frac{x-2}{x+3}\right| + \frac{5}{x+3} + C$

2 $\ln\left|\frac{x+1}{\sqrt{x^2+4}}\right| + \frac{1}{2}\tan^{-1}\frac{x}{2} + C$

3 a $\ln\left(\frac{1}{3}\right) - \frac{1}{6}$ b $\ln\sqrt{2} + \tan^{-1}2 - \tan^{-1}3$

c $\ln 3 + \pi$ d $2\ln\frac{3}{2} + \frac{1}{3}$

e $\ln\frac{32}{3} + 1$ f $\ln 2\sqrt{2} - \frac{\pi}{4}$

4 a $\ln\left|\frac{(x+1)(x-2)^4}{(x-1)^3}\right| + c$

b $\ln\left|\frac{x}{x-2}\right| - \frac{2}{x-2} + c$ c $\frac{1}{3}\ln\frac{20}{9}$

5 proof

6 a proof

b $\frac{1}{3}x^3 + x^2 + 4x + 8\ln|x-2| + c$

7 proof

Exercise 3B (page 67)

1 a $x^2 + 3x + \ln|x^2-1| + c$

b $\frac{3}{2}x^2 + \ln|x+1| + \frac{1}{\sqrt{2}}\tan^{-1}\frac{x}{\sqrt{2}} + c$

c $\frac{1}{2}x^2 + 5x + \frac{1}{2}\ln\left|\frac{x-3}{x+1}\right| + \frac{1}{x+1} + c$

d $6x + \ln|x-1| + \frac{1}{\sqrt{3}}\tan^{-1}\frac{x}{\sqrt{3}} + c$

e $\frac{1}{2}x^2 + x + \ln\left|\frac{(x-1)(x+2)}{x+3}\right| + c$

f $\frac{1}{3}x^3 - x^2 + 3x - 4\ln|x+1| - \frac{1}{x+1} + c$

2 proof

3 proof

4 $\frac{\pi}{3}$

5 0.896 units^2

Exercise 4 (page 69)

1 a $x\sin x + \cos x + c$

b $x\tan x + \ln|x\tan x| + c$

c $\left(\frac{1}{2}x^2 + 2x\right)\ln|x| - \frac{1}{4}x^2 - 2x + c$

d $-\cos x\ln|\cos x| + \cos x + c$

e $-\frac{1}{2x^2}\ln|x| - \frac{1}{4x^2} + c$

f $\frac{2}{9}x(1+3x)^{\frac{3}{2}} - \frac{4}{135}(1+3x)^{\frac{5}{2}} + c$

2 a $4\ln 2 - \frac{5}{4}$ b $-\frac{2}{9}$

c $-\frac{15}{14} + \frac{9}{7}\sqrt[3]{2}$ d $\frac{1}{8} - \frac{\pi\sqrt{3}}{48}$

e $\frac{1}{4}(e^2+1)$ f $-3e + 5$

3 $x\ln|x| - x + c$

4 b $\frac{1}{2}x^2\tan^{-1}x - \frac{1}{2}x + \frac{1}{2}\tan^{-1}x + c$

5 $\frac{\pi^2}{4}$

6 c $P = \frac{1}{2}e^x(\sin x + \cos x) + c$;

$Q = \frac{1}{2}e^x(\sin x - \cos x) + c$

Exercise 5A (page 71)

1 a $\frac{1}{2}x^2\sin 2x + \frac{1}{2}x\cos 2x - \frac{1}{4}\sin 2x + c$

b $e^{-x}(x^2 + 4x + 5) + c$

c $-(x^2+3)\cos x + 2x\sin x + 2\cos x + c$

d $e^x(x^3 - 3x^2 + 6x - 6) + c$

e $\frac{1}{2}e^x(\cos x + \sin x) + c$

f $\frac{1}{5}e^x(\sin 2x - 2\cos 2x) + c$

g $\frac{1}{2}x^2\ln x - \frac{1}{4}x^2 + c$

h $\frac{2}{3}x(x-1)^{\frac{3}{2}} - \frac{4}{15}(x-1)^{\frac{5}{2}} + c$

i $\frac{1}{11}x(x+4)^{11} - \frac{1}{132}(x+4)^{12} + c$

j $-\frac{1}{4}x(x+1)^{-4} + \frac{1}{12}(x+1)^{-3} + c$

k $-\frac{1}{12}x^2(2x-3)^{-6} - \frac{1}{60}x(2x-3)^{-5} + \frac{1}{480}(2x-3)^{-4} + c$

l $4x^2(x+3)^{\frac{1}{2}} - \frac{16}{3}x(x+3)^{\frac{3}{2}} + \frac{32}{15}(x+3)^{\frac{5}{2}} + c$

m $\frac{1}{2}x^2\ln x^3 - \frac{3}{4}x^2 + c$

n $-\frac{1}{3}x^2\cos(3x-1) - \frac{2}{9}x\sin(3x-1) - \frac{2}{27}\cos(3x-1) + c$

o $-\frac{1}{24}(5\sin x\cos 5x - \cos x\sin 5x) + c$

p $\frac{1}{16}(3\cos 3x\cos 5x + 5\sin 3x\sin 5x) + c$

q $\frac{1}{2}e^{-x}(-\sin x - \cos x) + c$

r $\frac{3^x}{(\ln 3)^2}(x\ln 3 - 1) + c$

s $-\frac{1}{5}e^{-x}(\sin^2 x + \sin 2x + 2) + c$

t $-x\cot x + \ln(\sin x) + c$

u $x\tan^{-1}x - \frac{1}{2}\ln|1 + x^2| + c$

2 a $\frac{\pi}{8} + \frac{1}{2}$ b $\frac{\pi}{48}(\pi^2 - 6)$

c $e - 2$

d $\frac{\pi}{2\sqrt{2}} - \ln(1 + \sqrt{2})$

e $\frac{16}{3}\ln 4 - \frac{28}{9}$ f 1

g $\pi\left(\frac{2\sqrt{3} - 3}{12}\right) + \ln\sqrt{2}$

h $\frac{1}{\sqrt{2}} + \frac{1}{2}\ln(1 + \sqrt{2})$

i $\frac{1}{2}\left(1 + \ln 2\right)^2$ j $\frac{1}{12}$

k $\frac{481}{120}$ l $\frac{20}{3}$

m 4 n $-\frac{\pi(\pi + 2)}{64\sqrt{2}}$

o 0

p $\frac{e^{-2}}{5}(\sin 1 - 2\cos 1) + \frac{2}{5}$

q $\frac{1}{(\ln 2)^3}(2(\ln 2)^2 - 4\ln 2 + 2)$

r 0

3 $\frac{1}{2}\left(\pi^2 - 4\right)$

Exercise 5B (page 72)

1 a proof

b $\frac{1}{2}(x^2 - 1)\ln(1 + x) - \frac{1}{4}(x - 1)^2 + c$

2 a proof b $\left(x - \frac{4}{3}\right)\ln(4 - 3x) - x + c$

3 a $e - 2$ b $\frac{3}{8} - \frac{19}{8e^2}$

4 $\sqrt{3}\ln 4 - 2\sqrt{3} + \frac{2\pi}{3}$

5 $\frac{1}{2}$

6 $2(\sqrt{x}\sin\sqrt{x} + \cos\sqrt{x}) + c$

7 33.6 units^3

Exercise 6 (page 74)

1 a $y = x^2 + c$ b $y = \frac{1}{2}ax^2 + c$

c $y = \frac{1}{3}x^3 - x + c$

d $y = \frac{1}{3}x^3 - \frac{1}{2}x^2 - 2x + c$

e $y = \frac{1}{3}\tan 3x + c$

f $y = \sin^{-1}\left(\frac{x}{a}\right) + c$

g $y = \frac{1}{3}\ln k\,|3x + 2|$

h $y = \frac{1}{2}x + \frac{1}{4}\sin 2x + c$

i $y = \frac{1}{4}\ln|k(\sin x)|$

j $y = -\frac{1}{12}(3x - 7)^{-4} + c$

k $y = \frac{1}{4}\tan^{-1}\left(\frac{x}{4}\right) + c$

l $y = \frac{1}{8}\tan^{-1}\left(\frac{x}{2}\right) + c$

2 a $y = \frac{1}{4}x^4 - 3$

b $y = -\frac{1}{n}\cos nx + 5 + \frac{1}{n}(-1)^n$

c $y = \frac{1}{2}\tan^{-1}\left(\frac{x}{2}\right) + \frac{\pi}{24}$

d $y = 2\ln|x - 1| - \frac{3}{2}x^2 + 6$

e $y = \frac{1}{3}e^{3x+2} + \frac{2}{3}e^2$

f $y = \frac{1}{4}\tan 4x + \frac{3}{4}$

3 $y = \ln\left|2\left(\frac{x - 1}{x}\right)\right|$

4 $y = \frac{1}{2}xe^{2x} - \frac{1}{4}e^{2x} - \frac{1}{4}e^2$

5 proof

6 proof

Exercise 7 (page 76)

1 a $y = ke^{2x}$ b $y = \sqrt{\frac{-1}{2x + c}}$

c $y = ke^{ax}$ d $y = \frac{1 + ke^{2x}}{1 - ke^{2x}}$

e $y = 3\sin(c - x)$

f $y = \sqrt{3}\tan\sqrt{3}(c + x)$

g $y = -\frac{1}{3}\ln(-3x + c)$ h $y = \frac{2 - ke^x}{1 - ke^x}$

2 a $y = 3(1 - e^{-x})$ b $\frac{y = 8 + e^{3x}}{4 - e^{3x}}$

c $y^2 = (x - 1)^2 + 1$ d $y = \tan^{-1}x$

e $y = \sqrt{3}\tan\left(\sqrt{3}x + \frac{\pi}{6}\right)$

f $y = \frac{1}{2}\left(\sqrt[4]{\frac{1}{9 - 8x}} + 3\right)$

Exercise 8 (page 77)

1 a $y = kx$ b $y^2 = \frac{4}{3}x^2 + c$

c $y = k(x + 1)^2$ d $y = \frac{-x}{1 + cx}$

e $y + 2 = k(x + 3)^{\frac{3}{2}}$

f $y = ke^{\sin^{-1}\left(\frac{x}{\sqrt{3}}\right)}$

g $y = 1 + \dfrac{cx}{1 + x}$

h $y + \frac{1}{2}\sin 2y = -2e^x + c$

i $y = \tan^{-1}(e^{-x} + c)$

2 a $2y(y + 1) = x^3 + x + 2$

b $y = \sin^{-1}\left(e^{-x} - \frac{1}{2}\right)$

c $y = \dfrac{1}{2 - \ln x - x}$

3 a $-xe^{-x} - e^{-x} + c$ **b** $y = \dfrac{e^x}{x + 1}$

4 a $\ln|\sin t| + c$

b $y = \tan\left(\ln|\sin x| + \dfrac{\pi}{4}\right)$

5 a $\dfrac{1}{t - 2} - \dfrac{1}{t - 1}$ **b** $y = \dfrac{4 - e^{\frac{1}{2}x^2}}{2 - e^{\frac{1}{2}x^2}}$

6 a $\ln|u| + c = \ln|\ln t| + c$ **b** $y = e^{2x}$

7 a proof

b $\ln|\operatorname{cosec} y - \cot y| = e^{-x} - 1$

8 $2y + 1 = \sqrt{3}\tan\left(\dfrac{\sqrt{3}}{4}x^2 + c\right)$

9 proof

10 proof

Exercise 9A (page 81)

1 a $\dfrac{ds}{dt} = ks$ **b** $\dfrac{ds}{dt} = \dfrac{k}{t}$

c $\dfrac{dn}{dt} = kn$ **d** $\dfrac{dV}{dt} = kr$

e $\dfrac{dN}{dt} = k(500 - N)$ **f** $\dfrac{dB}{dC} = \dfrac{k}{C}$

2 a $D = k\sqrt{h} + c$ **b** $D = 4.7\sqrt{h} + 8$

c 12.31 km

3 a $T = -\frac{1}{5}t^2 + \frac{3}{2}t + 98.4$ **b** $7\frac{1}{2}$ days

4 a $V = -\dfrac{10k}{3}e^{-0.3t} + c$ **b** $c = \dfrac{10k}{3}$

c $k = 29.7$, $V = -99e^{-0.3t} + 99$

d $V = 94$; model is overestimating V

5 $n = 3e^{1.5t}$

6 a $\dfrac{dV}{dt} = kV$ **b** $V = 20\,000e^{-0.067t}$

c (i) 10.3 years **(ii)** £10 234

7 a proof **b** $c = -6.215$

c $k = 0.026$ **d** 88 days

8 a $F = 400\left(1 - e^{-0.196t}\right)$ **b** $t = 8.2$ days

9 a proof **b** $c = -4 \times 10^{-3}$

c (i) $t = 12.7 + 1.7\ln\left(\dfrac{P}{2000 - P}\right)$

(ii) $P = \dfrac{2000e^{(0.6t - 8)}}{1 + e^{(0.6t - 8)}}$

d 13 days

10 a $P = \dfrac{100e^{(t - 4.6)}}{1 + e^{(t - 4.6)}}$ **b** 6 days

Exercise 9B (page 83)

1 18.5 °C

2 1.22 °C

3 a $\dfrac{dy}{dx} = 4$

b $y = 4x + c$; family of straight lines of gradient 4

4 a family of circles, centre the origin

b family of circles, centre (a, b)

5 385 hours

6 $h + b\ln h = -at + c$

7 24.2 g

8 9500 millions

9 a $v = -gt$ **b** $x = x_0 - \frac{1}{2}gt^2$

10 a £53 083.89 **b** £7.93

11 b 5.5 million

12 a $V = (10 - 0.2t)^2$ **b** 50 hours

13 $v = 5e^{-0.2t + 2.3}$; $449 - 25e^{-0.2t + 2.3}$

14 proof

Review (page 86)

1 a $\dfrac{\pi}{3}$ **b** $\frac{1}{5}\tan^{-1}\frac{2}{5}$ **c** $\dfrac{2\pi}{54}$

2 a $\frac{1}{5}\ln\left(\dfrac{x - 3}{x + 2}\right) + c$

b $\frac{1}{5}[\ln(x + 2) + 2\tan^{-1}x] + c$

c $\frac{2}{3}\ln\left(\dfrac{(x - 1)(x + 2)^2}{(x + 1)^3}\right) + c$

d $\frac{3}{10}\ln\left|\dfrac{x^2 + x + 5}{(x + 1)^2}\right| + \frac{47}{20}\tan^{-1}\left(\dfrac{2x + 1}{4}\right) + c$

e $x + \frac{2}{3}\ln\left(\dfrac{(x - 2)^5}{x + 1}\right) + c$

3 a $\frac{1}{2}e^{2x}(4x - 5) + c$ **b** $\frac{3}{8}(\pi + 2)$

4 a $y = ke^{x^2 + x}$ **b** $y = \sin^{-1}(x + c)$

5 $y = \dfrac{2}{3 - 4e^{2x}}$

6 70 °C

7 $\dfrac{dy}{dx} = -\dfrac{3x}{\sqrt{1 - x^2}}$; $x^2 + \frac{1}{9}\left(y - c\right)^2 = 1$; family of ellipses, centre $(0, c)$

CHAPTER 4

Exercise 1 (page 90)

1 **a** $5+5i$ **b** $2+11i$ **c** $6+3i$
d $6+8i$ **e** $17+16i$ **f** $3+4i$
g $2+11i$ **h** $-38+41i$ **i** $-3-4i$
j $-1-3i$ **k** $1+3i$ **l** $-3-i$

2 **a** $4+5i$ **b** 10 **c** 2
d $7-i$ **e** $7-24i$ **f** $2-10i$
g $6+2i$ **h** $2+16i$ **i** $30-10i$

3 **a** $z=-1\pm i$ **b** $-2\pm 3i$ **c** $3\pm 2i$
d $1\pm 2i$ **e** $2\pm i$ **f** $-3\pm 3i$

4 proofs

5 $5\pm\sqrt{15}i$

6 **a** (i) 10 (ii) 13 (iii) 5
b all real **c** a^2+b^2

7 **a** $i, -1, -i, 1, i, -1, -i, 1, i, -1, -i, 1$
b (i) $-i$ (ii) i (iii) -1 (iv) 1 (v) $-i$

8 **a** $a=8, b=6$ **b** $a=5, b=12$
c $a=2, b=11$

Exercise 2 (page 91)

1 **a** $2-2i$ **b** $2-i$ **c** $1+i$
d $1-i$ **e** $1.6-3.2i$
f $1.7+0.9i$

2 **a** $-i$ **b** $\frac{1}{2}+\frac{1}{2}i$ **c** $\frac{1}{4}-\frac{1}{4}i$
d $\frac{3}{10}-\frac{1}{10}i$ **e** $\frac{1}{5}+\frac{1}{10}i$

3 **a** $3-2i$ **b** $3-3i$ **c** $2-5i$
d $-\frac{3}{2}-\frac{7}{2}i$ **e** $-0.2-1.6i$
f $0.36-0.48i$

4 **a** $3, -2$ or $-3, 2$ **b** $4, -1$ or $-4, 1$
c $1, -5$ or $-1, 5$

5 **a** $2-i$ or $-2+i$ **b** $5-2i$ or $-5+2i$
c $4+5i$ or $-4-5i$

6 **a** (i) $2-3i$ (ii) $\frac{2}{13}+\frac{3}{13}i$
(iii) $-\frac{5}{13}+\frac{12}{13}i$ (iv) $-\frac{5}{13}-\frac{12}{13}i$
(v) $-\frac{10}{13}$ (vi) $\frac{24}{13}i$

b (i) $a-ib$ (ii) $\frac{a}{a^2+b^2}+\frac{b}{a^2+b^2}i$
(iii) $\frac{a^2-b^2}{a^2+b^2}+\frac{2ab}{a^2+b^2}i$
(iv) $\frac{a^2-b^2}{a^2+b^2}-\frac{2ab}{a^2+b^2}i$
(v) $\frac{2(a^2-b^2)}{a^2+b^2}$ (vi) $\frac{4ab}{a^2+b^2}i$

7 **b** $\mathcal{I}(z)=\frac{1}{2i}(z+\bar{z})$

8 **a** (i) $a-bi$ (ii) $x-yi$
(iii) $(a+x)-(b+y)i$
b $\bar{z}_1+\bar{z}_2=\overline{z_1+z_2}$
c (i) $\bar{z}_1-\bar{z}_2$ (ii) $\bar{z}_1\times\bar{z}_2$ (iii) $\bar{z}_1\div\bar{z}_2$

Exercise 3 (page 94)

1 **a** $|z|=1$; $\arg(z)=\frac{\pi}{2}$
b $|z|=1$; $\arg(z)=\pi$
c $|z|=1$; $\arg(z)=-\frac{\pi}{2}$

2 **a** (i) $(3, 4), (3, -4)$ (ii) $(2, 3), (2, -3)$
(iii) $(5, 1), (5, -1)$
b reflection in x-axis

3 **a** (i) $(1, 1)$ (ii) $1.41, 45°$
b (i) $(2, 3)$ (ii) $3.61, 56.3°$
c (i) $(3, 2)$ (ii) $3.61, 33.7°$
d (i) $(6, 0)$ (ii) $6, 0°$
e (i) $(0, 3)$ (ii) $3, 90°$
f (i) $(-4, -3)$ (ii) $5, -143°$
g (i) $(-1, 2)$ (ii) $2.24, 117°$
h (i) $(2, -3)$ (ii) $3.61, -56.3°$
i (i) $(4, -1)$ (ii) $4.12, -14.0°$

4 **a** (i) $\frac{1}{2}-\frac{1}{2}i$ (ii) $0.707, -45°$
b (i) $\frac{1}{10}-\frac{3}{10}i$ (ii) $0.316, -71.6°$
c (i) $6+2i$ (ii) $6.32, 18.4°$

5 **a** (i) $12.2, 35°$ (ii) $149, 70°$
(iii) $1819, 105°$
b Raise a number to the nth power and you raise the modulus by the nth power and increase the argument by a factor of n.

6 **a** $\sqrt{3}+i$ **b** $\frac{3}{\sqrt{2}}+\frac{3}{\sqrt{2}}i$ **c** $4i$
d $\frac{3}{2}+\frac{3\sqrt{3}}{2}i$ **e** $\sqrt{2}-\sqrt{2}i$ **f** $\frac{\sqrt{3}}{2}-\frac{1}{2}i$

7 **a** $2(\cos 60°+i\sin 60°)$
b $2(\cos 45°+i\sin 45°)$
c $4(\cos 150°+i\sin 150°)$
d $(\cos 180°+i\sin 180°)$
e $3(\cos 90°+i\sin 90°)$
f $8(\cos 120°-i\sin 120°)$
g $4(\cos 150°-i\sin 150°)$
h $2(\cos 135°-i\sin 135°)$
i $2(\cos 120°-i\sin 120°)$

8 diagrams to illustrate
a $(5, 5)$ **b** $(5, 5)$ **c** $(5, -5)$
d $(-5, 1)$ **e** $(-1, 1)$ **f** $(-1, -1)$
g $(-2, 3)$ **h** $(-5, 1)$ **i** $(1, 1)$

Exercise 4 (page 96)

1 **a** $x^2+y^2=25$ **b** $(x-3)^2+y^2=4$
c $(x+1)^2+y^2=16$ **d** $x^2+(y+1)^2=9$
e $x^2+(y-2)^2=9$
f $(x+1)^2+(y+2)^2=9$
g $(2x)^2+(2y+3)^2=25$
h $(3x)^2+(3y-1)^2=25$
i $(3x+3)^2+(3y-2)^2=16$
j $y=\frac{x}{\sqrt{3}}$ **k** $y=x$

l $y = -\sqrt{3}x$ **m** $y = x \tan 1$

n $y = \tan \frac{\pi}{8}x$ **o** $y = -\sqrt{3}x$

2 **a** circle, centre $(a, 0)$, radius b
b circle, centre $(0, a)$, radius b
c circle, centre (b, a), radius c

3 **a** $x = y$ **b** $y = 2x - \frac{3}{2}$

c $y = \frac{3}{2}x - \frac{5}{4}$ **d** $\frac{a}{b}x - \frac{a^2 - b^2}{2b} = y$

4 **a**

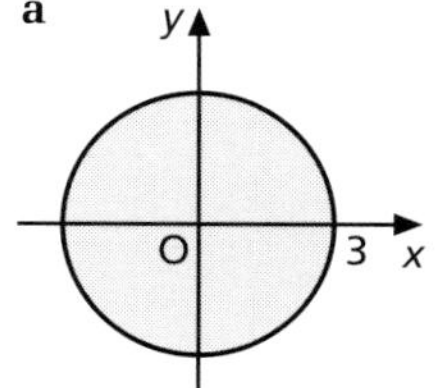

b

c

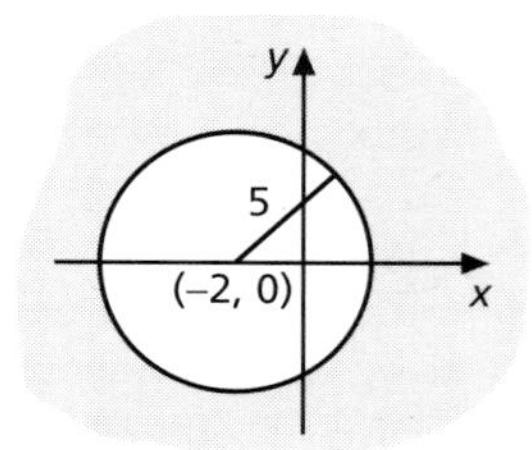

Note. In order to save space, in the answers for the rest of this chapter an abbreviation cis is used as follows: $a \text{ cis } x = a(\cos x + i \sin x)$, e.g.

$$3 \text{ cis } \frac{\pi}{4} = 3\left(\cos \frac{\pi}{4} + i \sin \frac{\pi}{4}\right),$$
$$4 \text{ cis } (-20°) = 4(\cos(-20°) + i \sin(-20°))$$
$$= 4(\cos 20° - i \sin 20°).$$

Exercise 5 (page 98)

1 **a** $12 \text{ cis } \frac{5\pi}{6}$ **b** $10 \text{ cis } \frac{5\pi}{12}$

c $8 \text{ cis } 0 = 8$ **d** $2 \text{ cis } \left(-\frac{5\pi}{6}\right)$

e $10 \text{ cis } \left(-\frac{\pi}{30}\right)$ **f** $2 \text{ cis } \frac{\pi}{6}$

g $\frac{5}{2} \text{ cis } \frac{\pi}{8}$ **h** $3 \text{ cis } \frac{2\pi}{3}$

i $4 \text{ cis } \frac{\pi}{7}$

2 **a** $z_1 = 5 \text{ cis } (0.927)$,
$z_2 = \sqrt{2} \text{ cis } (0.785)$,
$z_1 z_2 = 7.07 \text{ cis } (1.712)$,
$\frac{z_1}{z_2} = 3.54 \text{ cis } (0.142)$

b $z_1 = 3.61 \text{ cis } (0.983)$,
$z_2 = 3.16 \text{ cis } (-0.321)$,
$z_1 z_2 = 11.4 \text{ cis } (0.661)$,
$\frac{z_1}{z_2} = 1.14 \text{ cis } (1.30)$

c $z_1 = 1.414 \text{ cis } (-0.785)$,
$z_2 = 1.414 \text{ cis } (-2.35)$,
$z_1 z_2 = 2 \text{ cis } (\pi)$,
$\frac{z_1}{z_2} = 1 \text{ cis } \left(\frac{\pi}{2}\right)$

d $z_1 = 4.12 \text{ cis } (2.90)$,
$z_2 = 2.83 \text{ cis } (2.36)$,
$z_1 z_2 = 11.7 \text{ cis } (-1.03)$,
$\frac{z_1}{z_2} = 1.46 \text{ cis } (0.540)$

3 **a** $r^2 \text{ cis } \frac{2\pi}{3}$ **b** $r^3 \text{ cis } \pi$

c $r^4 \text{ cis } \frac{-2\pi}{3}$ **d** $r^5 \text{ cis } \frac{-\pi}{3}$

e $r^6 \text{ cis } 0$ **f** $r^7 \text{ cis } \frac{\pi}{3}$

4 **(i)** **a** $r^2 \text{ cis } \pi$ **b** $r^3 \text{ cis } \frac{-\pi}{2}$

c $r^4 \text{ cis } 0$ **d** $r^5 \text{ cis } \frac{\pi}{2}$

e $r^6 \text{ cis } \pi$ **f** $r^7 \text{ cis } \frac{-\pi}{2}$

(ii) **a** $r^2 \text{ cis } \frac{-2\pi}{3}$ **b** $r^3 \text{ cis } 0$

c $r^4 \text{ cis } \frac{2\pi}{3}$ **d** $r^5 \text{ cis } \frac{-2\pi}{3}$

e $r^6 \text{ cis } 0$ **f** $r^7 \text{ cis } \frac{2\pi}{3}$

(iii) **a** $r^2 \text{ cis } \frac{-2\pi}{4}$ **b** $r^3 \text{ cis } \frac{\pi}{4}$

c $r^4 \text{ cis } \pi$ **d** $r^5 \text{ cis } \frac{-\pi}{4}$

e $r^6 \text{ cis } \frac{2\pi}{4}$ **f** $r^7 \text{ cis } \frac{-3\pi}{4}$

(iv) **a** $r^2 \text{ cis } 2\theta$ **b** $r^3 \text{ cis } 3\theta$
c $r^4 \text{ cis } 4\theta$ **d** $r^5 \text{ cis } 5\theta$
e $r^6 \text{ cis } 6\theta$ **f** $r^7 \text{ cis } 7\theta$

5 **a** $-527 - 336i$
b **(i)** $5 \text{ cis } (0.927)$
(ii) $625 \text{ cis } (-2.575)$
(iii) $-527 - 336i$
c method **b**

Exercise 6 (page 101)

1 **a** **(i)** $8 + 13.9i$ **(ii)** $-887 + 512i$
(iii) $524\,288 - 908\,093i$
b **(i)** $-8i$ **(ii)** $-8 - 13.9i$
(iii) $-128 + 221.7i$
c **(i)** $-2 + 2i$ **(ii)** $-8i$
(iii) -64

2 **a** $-8.34 + 25.6i$ **b** $-128 - 222i$
c 1 **d** $-0.223 + 0.975i$

3 **a** $-46 + 9i$ **b** $-1121 + 404i$
c $-119 - 120i$ **d** $1024 + 1024i$

4 **a** cis $40°$ **b** cis $(-165)°$
c cis $5°$ **d** cis $170°$
e cis $20°$ **f** cis $(-6)°$
g cis $120°$ **h** cis $175°$
i cis $100°$ **j** cis $24°$
k cis $90°$ **l** cis $88°$

5 **a** **(i)** $\cos^2\theta + 2i\cos\theta\sin\theta - \sin^2\theta$
(ii) $\cos 2\theta + i\sin 2\theta$
b **(i)** $\cos 2\theta = \cos^2\theta - \sin^2\theta$
(ii) $\sin 2\theta = 2\sin\theta\cos\theta$

6 **a** **(i)** $\cos^3\theta + 3i\cos^2\theta\sin\theta - 3\cos\theta\sin^2\theta - i\sin^3\theta$
(ii) $\cos 3\theta + i\sin 3\theta$
b **(i)** $\cos 3\theta = \cos^3\theta - 3\cos\theta\sin^2\theta$
(ii) $\cos 3\theta = 4\cos^3\theta - 3\cos\theta$
c $\sin 3\theta = 3\cos^2\theta\sin\theta - \sin^3\theta = 3\sin\theta - 4\sin^3\theta$
d $\sin^3\theta = \frac{1}{4}(3\sin\theta - \sin 3\theta)$

7 **a** **(i)** $\cos 4\theta = 8\cos^4\theta - 8\cos^2\theta + 1$
(ii) $\sin 4\theta = 4\cos^3\theta\sin\theta - 4\cos\theta\sin^3\theta$
(iii) $\cos^4\theta = \frac{1}{8}(\cos 4\theta + 8\cos^2\theta - 1)$
b **(i)** $\cos 5\theta = 16\cos^5\theta - 20\cos^3\theta + 5\cos\theta$
(ii) $\sin 5\theta = 5\sin\theta - 20\sin^3\theta + 16\sin^5\theta$
(iii) $\cos^5\theta = \frac{1}{16}(\cos 5\theta - 5\cos\theta + 20\cos^3\theta)$

8 **a** **(i)** cis $\left(-\frac{11\pi}{3}\right)$ **(ii)** cis $\left(-\frac{5\pi}{3}\right)$
(iii) cis $\left(\frac{\pi}{3}\right)$ **(iv)** cis $\left(\frac{7\pi}{3}\right)$
b **(i)** $\frac{\pi}{3}, \frac{\pi}{3}, \frac{\pi}{3}, \frac{\pi}{3}$
(ii) one distinct answer
c **(i)** $\frac{\pi}{6}, -\frac{5\pi}{6}, \frac{\pi}{6}, -\frac{5\pi}{6}$
(ii) cis $\left(\frac{\pi}{6}\right)$, cis $\left(-\frac{5\pi}{6}\right)$
(iii)

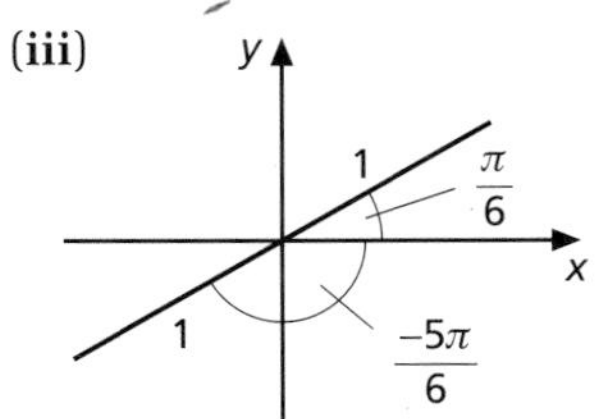

9 **a** proofs
b **(i)** $\frac{11\pi}{12}, -\frac{5\pi}{12}, \frac{\pi}{4}, \frac{11\pi}{12}, -\frac{5\pi}{12}$
(ii) cis $\left(\frac{11\pi}{12}\right)$, cis $\left(-\frac{5\pi}{12}\right)$, cis $\left(\frac{\pi}{4}\right)$
(iii)

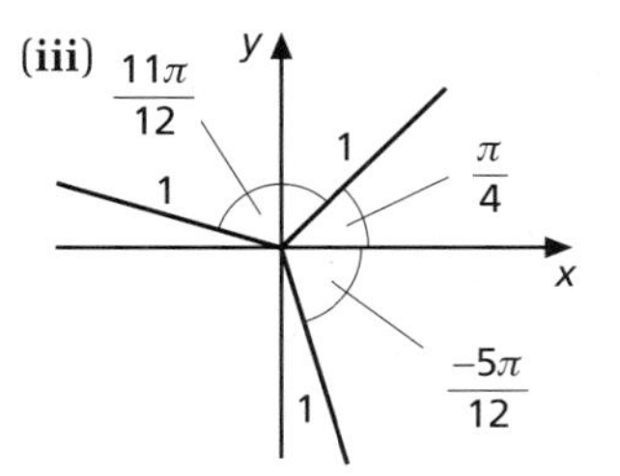

Exercise 7 (page 106)

1 **a** $2\text{ cis}\left(\frac{1}{3}\left(\frac{\pi}{4} + 2k\pi\right)\right), k = 0, 1, 2$
b $\text{cis}\left(\frac{1}{4}\left(\frac{\pi}{5} + 2k\pi\right)\right), k = 0, 1, 2, 3$
c $2\text{ cis}\left(\frac{1}{5}\left(\frac{\pi}{7} + 2k\pi\right)\right), k = 0, \ldots, 4$
d $4\text{ cis}\left(\frac{1}{3}\left(\frac{2\pi}{3} + 2k\pi\right)\right), k = 0, 1, 2$
e $2\text{ cis}\left(\frac{1}{5}\left(-\frac{\pi}{7} + 2k\pi\right)\right), k = 0, \ldots, 4$
f $4\text{ cis}\left(\frac{1}{3}\left(-\frac{2\pi}{3} + 2k\pi\right)\right), k = 0, 1, 2$
g $2^{\frac{3}{8}}\text{ cis}\left(\frac{1}{4}\left(-\frac{3\pi}{4} + 2k\pi\right)\right), k = 0, 1, 2, 3$
h $6^{\frac{1}{5}}\text{ cis}\left(\frac{1}{5}\left(\frac{2\pi}{3} + 2k\pi\right)\right), k = 0, \ldots, 4$
i $8^{\frac{1}{8}}\text{ cis}\left(\frac{1}{4}\left(\frac{3\pi}{4} + 2k\pi\right)\right), k = 0, 1, 2, 3$
j $6^{\frac{1}{5}}\text{ cis}\left(\frac{1}{5}\left(-\frac{2\pi}{3} + 2k\pi\right)\right), k = 0, \ldots, 4$

2 **a** $\text{cis}\left(\frac{1}{3}(2k\pi)\right)$ **b** $\text{cis}\left(\frac{1}{4}(2k\pi)\right)$
c $\text{cis}\left(\frac{1}{6}(2k\pi)\right)$ **d** $3\text{ cis}\left(\frac{1}{4}(2k\pi)\right)$
e **(i)** $\text{cis}\left(\frac{1}{5}(\pi + 2k\pi)\right)$ **(ii)** $\left(\frac{1}{5}\left(\frac{\pi}{2} + 2k\pi\right)\right)$
(iii) $\text{cis}\left(\frac{1}{5}\left(-\frac{\pi}{2} + 2k\pi\right)\right)$
f **(i)** $4\text{ cis}\left(\frac{1}{3}(\pi + 2k\pi)\right)$
(ii) $5\text{ cis}\left(\frac{1}{4}\left(\frac{\pi}{2} + 2k\pi\right)\right)$
(iii) $\frac{1}{2}\text{ cis}\left(\frac{1}{5}\left(-\frac{\pi}{2} + 2k\pi\right)\right)$

Exercise 8 (page 108)

1 **a** $1 \pm 3i$ **b** $2 \pm i$ **c** $3 \pm 4i$
d $2 \pm 0.5i$ **e** $-\frac{1}{2} \pm \frac{1}{2}i$
f $-0.4 \pm 1.2i$

2 **a** $\pm i$ **b** $-1 \pm 2i$ **c** $-2 \pm 5i$
d $-2 \pm i$ **e** $-0.5 \pm 1.5i$ **f** $-3 \pm 2i$

3 **a** $3, -2 \pm i$ **b** $4, 1 \pm 2i$
c $-2.5, \pm 2i$ **d** $-1.5, -1 \pm 3i$
e $\pm 1.5i, 1$ **f** $-2 \pm 3i, -2.5$

4 $(z+2)(z-1), z^2 - 2z + 2$

5 $(z-2)(z-1), z^2 - 2z + 5$

6 **a** $2 - i, -1 \pm i,$ **b** $3 - 2i, 1 \pm 2i$
c $1 - 3i, -1, 0.5$

7 **a** $4 \pm i, \pm 2$ **b** $(2 \pm 3i), \frac{2}{3}, \frac{1}{2}$
c $(4 \pm 2i), 5, -\frac{1}{2}$

8 $-2 \pm i, -1 \pm i, -1$

Review (page 110)

1 **a** $8 - 8i$ **b** $2 - 16i$
c $63 - 16i$ **d** $5 + 12i$
e $-1.32 - 2.24i$ **f** $3 - 2i$ or $-3 + 2i$
g 3

2 $a = 2, b = 1$

3 $13(\cos 67.4^\circ + i \sin 67.4^\circ)$

4 $3.54, -81.9^\circ$ (3 s.f.)

5

6 **a** $(x+1)^2 + y^2 = 25$
b

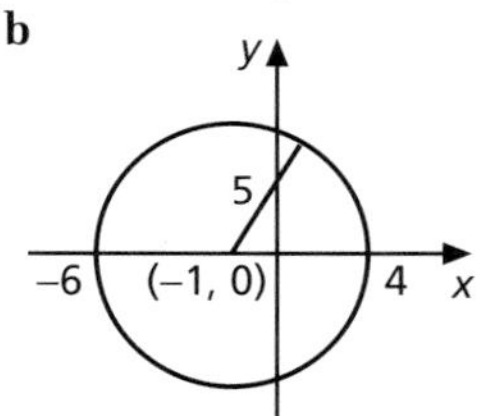

7 **a** $6 \text{ cis } \frac{\pi}{2}$ **b** $\frac{2}{3} \text{ cis } \frac{\pi}{6}$ **c** $8 \text{ cis } \pi$

8 **a** $256 \text{ cis } (-120^\circ)$ **b** $-128 - 221.7i$

9 **a** (i) $\cos^5\theta + 5i\cos^4\theta\sin\theta - 10\cos^3\theta\sin^2\theta - 10i\cos^2\theta\sin^3\theta + 5\cos\theta\sin^4\theta + i\sin^5\theta$
(ii) $\cos 5\theta + i \sin 5\theta$
b $\sin 5\theta = 5\sin\theta - 20\sin^3\theta + 16\sin^5\theta$

10 **a** $z = 5 \text{ cis } \left(\frac{1}{3}\left(\frac{\pi}{4} + 2k\pi\right)\right), k = 0, 1, 2$

11 $\text{cis } \left(\frac{2k\pi}{5}\right), k = 0, 1, 2, 3, 4$

12 $(z+2), (z-1), z^2 - 2z + 17$

13 $2 \pm 3i, 1 \pm 2i$

CHAPTER 5

Exercise 1 (page 115)

1 **a** (i) 8 (ii) $u_n + 2$
b (i) 11 (ii) $u_n + 3$
c (i) −3 (ii) $u_n - 6$
d (i) 24 (ii) $2u_n$
e (i) 4 (ii) $\frac{u_n}{2}$
f (i) 22 (ii) $2u_n + 2$
g (i) 17 (ii) $2u_n - 1$
h (i) 106 (ii) $4u_n + 2$
i (i) 5 (ii) $\frac{u_n}{2} - \frac{3}{2}$

2 **a** $u_{n+1} = 3u_n + 2$
b (i) 17 (ii) 485

3 **a** (i) $r^2u_1 + rd + d$ (ii) $r^3u_1 + r^2d + rd + d$
b (i) 3 (ii) −2

4 **a** −2, unstable **b** 20, stable
c 0.5, unstable **d** 10, stable
e undefined **f** −10, stable

5 **a** 20 million litres
b stable, at 30 million litres

Exercise 2A (page 117)

1 **a** $a = 3, d = 2$ **b** 4, 1 **c** 3, −2
d −2, 4 **e** −2, −3 **f** 0, 3
g 2, −0.1 **h** $\frac{1}{12}, \frac{1}{12}$ **i** $\frac{1}{8}, \frac{1}{16}$

2 **a** $3n + 1$ **b** $-3n + 11$ **c** $-4n + 13$
d $5n - 8$ **e** $-6n + 23$ **f** $\frac{1}{8}n + \frac{3}{8}$
g $\frac{1}{10}n + \frac{1}{5}$ **h** $-\frac{2}{9}n + \frac{10}{9}$
i $-0.7n + 0.1$

3 **a** 10 **b** 7 **c** 61

4 **a** 3 **b** −4
c $1, 1\frac{1}{3}, 1\frac{2}{3}, 2$

5 **a** 8 **b** 15 **c** 48

6 **a** (i) 4, 9, 14, 19 (ii) 25, 23, 21, 19
(iii) $1, 1\frac{3}{4}, 2\frac{1}{2}, 3\frac{1}{4}$
b (i) $a = 15, d = 7$ (ii) 155

7 **a** 3 **b** $x = \frac{q}{p}, x \in N$

8 $u_{x+1} - u_x = p$

9 **a** $\ln 6 - \ln 2 = \ln 18 - \ln 6 = \ln 3$
b $\ln(2 \times 3^{n-1})$ **c** 92

Exercise 2B (page 118)

1 **a** 1 **b** −11 **c** 29th

2 **a** 4, 8
b (i) proof (ii) 8, 8
c whenever the initial rectangle has one dimension equal to 2

3 **a** 17 **b** 5th

4 a 880
b potential common difference is not a whole number
5 common difference = π, $a = 20\pi$

Exercise 3A (page 120)

1 a 345 b 1350
c (i) 675 (ii) 20 500 (iii) −992 (iv) 7
2 a (i) 621 (ii) 684
b 63
c −243, −285, −42
3 a 324 b 9175 c −480
d −216 e 14.05 f 34
4 a 7 b 1.15
5 a 19
b 2, 3.8, 5.6, 7.4 and 1, $1\frac{3}{7}$, $1\frac{6}{7}$, $2\frac{2}{7}$
6 a 7 b 0.1
7 a 114 b −83
8 a 23, 212.5 b (i) −1 (ii) −11
9 10 100

Exercise 3B (page 121)

1 a $x = \dfrac{16}{(b-1)}$
b $(x, b) = (16, 2), (8, 3), (4, 5), (2, 9), (1, 17)$
c 520, 440, 400, 380, 370
2 a conditions lead to $x^3 - x = 3d \Rightarrow d = \dfrac{x(x-1)(x+1)}{3}$ and one of x, $(x-1)$ and $(x+1)$ is divisible by 3
b (i) $n^2 + n$, $4n^2 - n$, $10n^2 - 6n$, $20n^2 - 15n$
(ii) column 3 of triangle (see Bk 1 Ch 1)
3 a 4π, 4.02π, 4.04π, 4.06π, 4.08π
b 499π
4 a 7125 b £182 750
5 a 13 b 13

Exercise 4A (page 123)

1 a (i) 1, 3 (ii) $1 \times 3^{n-1}$
b (i) 4, −2 (ii) $4 \times (-2)^{n-1}$
c (i) 1458, $\frac{1}{3}$ (ii) $1458 \times \left(\frac{1}{3}\right)^{n-1}$
d (i) 3072, $-\frac{1}{4}$ (ii) $3072 \times \left(-\frac{1}{4}\right)^{n-1}$
e (i) 7, $\frac{1}{10}$ (ii) $7 \times \left(\frac{1}{10}\right)^{n-1}$
f (i) $\frac{2}{3}$, $\frac{2}{5}$ (ii) $\frac{2}{3} \times \left(\frac{2}{5}\right)^{n-1}$
g (i) 0.16, 0.8 (ii) $0.16 \times 0.8^{n-1}$
h (i) 23.2, 0.2 (ii) $23.2 \times 0.2^{n-1}$
2 a 2592 b 0.001 944
3 a $10 \times 5^{n-1}$ b 3 125 000
4 a 8 b 9
5 a 1.21 b 151
6 a d would need to be zero and hence $r = 1$
b (i) $a = \dfrac{d}{4}$ (ii) 5 c $x - 1$

7 a $\dfrac{\sin 2x}{2\sin x} = \dfrac{\sin 2x \cos x}{\sin 2x} = \cos x$
b $\sqrt{2}, 1, \dfrac{1}{\sqrt{2}}, \dfrac{1}{2}, \dfrac{1}{2\sqrt{2}}, \dfrac{1}{4}, \dfrac{1}{4\sqrt{2}}, \dfrac{1}{8}, \dfrac{1}{8\sqrt{2}}, \dfrac{1}{16}$
8 a (i) 12 (ii) $\dfrac{9}{10}$ b 250, 50

Exercise 4B (page 124)

1 a (i) 17 (ii) 27 (iii) 15
b (i) 17, 25 (ii) 27, 36 (iii) 17, 21
(iv) 18, 23
2 a (i) 2060 (ii) 2122 (iii) 2185
b $r = 1.03$ c 32
3 a $P\left(1 + \dfrac{r}{100}\right)$ b $P\left(1 + \dfrac{r}{100}\right)^2$
c $\left(1 + \dfrac{r}{100}\right)$ d $P\left(1 + \dfrac{r}{100}\right)^n$
4 a 7.7° b 46
5 a (i) $\dfrac{N_0e^{2a}}{N_0e^{a}} = \dfrac{N_0e^{3a}}{N_0e^{2a}}$ (ii) $r = e^a$
b 5
6 a (i) $\dfrac{V_0e^{\frac{-T}{5}}}{V_0e^{\frac{-T}{10}}} = \dfrac{V_0e^{\frac{-3T}{10}}}{V_0e^{\frac{-T}{5}}}$ (ii) $r = e^{\frac{-T}{10}}$
b (i) 6.9 years (ii) 23 years

Exercise 5A (page 127)

1 a 2186 b 2044 c −195 312
d 15 624 e −341 f 6554
2 a $\frac{91}{243}$ b 2.13 c 8.37
d 28.2 e 3.69 f 35.7
3 a 10 b 9
4 a $\frac{2}{3}$ b 170
5 a proof b 1, 3, 9, 27, 81
c 1, 2, 4, 8, 16
d (i) 11 625 (ii) 3, 15, 75, 375
6 a $\frac{1}{2}$ b 14 762
7 a $a = 1$, $r = \dfrac{1 \pm \sqrt{5}}{2}$ b (i) $\dfrac{1 - \phi^n}{1 - \phi}$
8 a (i) 1023, 1 048 575, 1 073 741 823
(ii) 1.25, 1.25, 1.25 (2 d.p.)
(iii) −682, −699 050, −715 827 882
(iv) 1.429, 1.429, 1.429 (3 d.p.)
(v) 130.26, 175.68, 191.52 (2 d.p.)
(vi) 571.62, 587.77, 588.22
b $|r| < 1$ each sequence tends to limit

Exercise 5B (page 128)

1 a 4.38 cm b 53.6 cm
2 a £188 355, £194 005, £199 825, £205 820, £211 995
b assume 1st yr constant £188 355 then increase is £20 191

3 **a** $57.6°$ **b** 10.8 seconds

4 **a** **(i)** 2 m **(ii)** $2\frac{11}{12}$ m **(iii)** 6.5 m

b **(i)** 9.75 m **(ii)** 1.5 times as big

5 **a** annual rate is 1.3%; $V_1 = 0.99V_0$; $V_2 = 0.97V_0$; $V_3 = 0.96V_0$; $V_4 = 0.95V_0$;

b 15% (nearest whole no.)

6 **a** **(i)** a **(ii)** $ra + d$ **(iii)** $r^2a + rd + d$ **(iv)** $r^3a + r^2d + rd + d$

b $u_n = ar^{n-1} + d\left(\frac{1 - r^{n-1}}{1 - r}\right)$

c $a\left(\frac{1 - r^n}{1 - r}\right) + d\left[\left(\frac{n - 1}{1 - r}\right) - \frac{r(1 - r^{n-1})}{(1 - r)^2}\right]$

Exercise 6A (page 131)

1 **a** 2 **b** 24 **c** $\frac{49}{6}$ **d** $\frac{16}{3}$ **e** 72.9 **f** $\frac{1000}{9}$

2 **a** 22.5 **c** -10 **e** $\frac{9}{5}$

3 **a** 32 **b** $r = \pm 0.4$, $\frac{90}{7}$ or 30

4 **a** $\frac{14}{100} + \frac{14}{10\,000} + \frac{14}{1\,000\,000} + \cdots = \frac{14}{99}$

b $\frac{1}{10} + \frac{4}{100} + \frac{4}{1000} + \frac{4}{10\,000} + \cdots = \frac{1}{10} + \frac{4}{90} = \frac{13}{90}$

c $\frac{270}{1000} + \frac{270}{1\,000\,000} + \cdots = \frac{270}{999} = \frac{10}{37}$

5 **a** $\frac{14}{990} = \frac{7}{495}$ **b** $\frac{608}{990} = \frac{304}{495}$

c **(i)** $\frac{1}{90}$ **(ii)** $\frac{29}{90}$ **(iii)** $\frac{432}{990}$ **(iv)** $\frac{4968}{9900}$

6 **a** 16 **b** 19.9936

7 **a** 6, 2.4, 0.96, 0.384

b **(i)** 9.8976 m **(ii)** 9.998 95 m

c 10 m

Exercise 6B (page 132)

1 **a** **(i)** $1010\frac{10}{99}$

(ii) Achilles and the tortoise are at the same spot $1010\frac{10}{99}$ m from the start

b **(i)** 100, 1, 0.01, 0.0001 **(ii)** $101\frac{1}{99}$ s

(iii) Achilles won't overtake the tortoise before $101\frac{1}{99}$ seconds are up.

2 **a** **(i)** 12 times **(ii)** 15 min **(iii)** $\frac{15}{12}$ min

(iv) $a = 15$, $r = \frac{1}{12}$; sum $= 16\frac{4}{11}$ min after 3

b $32\frac{8}{11}$ min after 6

Exercise 7A (page 134)

1 **a** **(i)** 12 **(ii)** 4 **(iii)** 94

b $1 + 2x + 3x^2 + 4x^3 + \cdots$; No

c $1 + 3x + 6x^2 + 10x^3 + \cdots$

2 $1 - x + x^2 - x^3 + \cdots$

3 **a** $-x - \frac{x^2}{2} - \frac{x^3}{3} - \frac{x^4}{4} - \cdots$; not geometric

b $x - \frac{x^2}{2} + \frac{x^3}{3} - \frac{x^4}{4} + \cdots$

c $2\left(x + \frac{x^3}{3} + \frac{x^5}{5} + \frac{x^7}{7} + \cdots\right)$

4 **a** $1 + x^2 + x^4 + x^6 + \cdots$ **b** even **c** $\sum_{r=0}^{\infty} x^{2r}$

5 **a** $1 + (1 - x) + (1 - x)^2 + (1 - x)^3 + \cdots$

b $-(1 - x) - \frac{(1 - x)^2}{2} - \frac{(1 - x)^3}{3} - \frac{(1 - x)^4}{4} - \cdots$

Exercise 7B (page 134)

1 **a** $1 - x^2 + x^4 - x^6 + \cdots$

b $x - \frac{x^3}{3} + \frac{x^5}{5} - \frac{x^7}{7} + \cdots$

c **(i)** $1 - \frac{1}{3} + \frac{1}{5} - \frac{1}{7} + \cdots = \frac{\pi}{4}$

2 **a** $1 + 2x + 4x^2 + 8x^3 + \cdots$; $|2x| < 1 \Rightarrow |x| < \frac{1}{2}$

b $1 - 3x + 9x^2 - 27x^3 + 81x^4 - \cdots$; $|x| < \frac{1}{3}$

c $1 + \frac{1}{x} + \frac{1}{x^2} + \frac{1}{x^3} + \cdots$; $|x| > 1$

3 **a** $1 + \sin x + \sin^2 x + \sin^3 x + \cdots$

b $1 + \sin^2 x + \sin^4 x + \sin^6 x + \cdots$

c $1 + \cos^2 x + \cos^4 x + \cos^6 x + \cdots$

4 **a** $\frac{1}{p} + \frac{q}{p^2} + \frac{q^2}{p^3} + \frac{q^3}{p^4} + \cdots$

b $\frac{1}{p} - \frac{q}{p^2} + \frac{q^2}{p^3} - \frac{q^3}{p^4} + \cdots$

5 **a** $\frac{1}{3} - \frac{4x}{9} + \frac{16x^2}{27} - \frac{64x^3}{81} + \cdots$; $-\frac{3}{4} < x < \frac{3}{4}$

b $\frac{1}{2} + \frac{3x}{4} + \frac{9x^2}{8} + \frac{27x^3}{16} + \cdots$; $-\frac{2}{3} < x < \frac{2}{3}$.

Each could have been expanded differently as our answer to **c** should illustrate:

c $\frac{1}{x - 2} = \frac{1}{x}.\frac{1}{1 - \frac{2}{x}} = \frac{1}{x} + \frac{2}{x^2} + \frac{4}{x^3} + \frac{8}{x^4} \cdots$;

$x > 2$ *or* $x < -2$ alternatively

$\frac{1}{x - 2} = \frac{1}{-2 + x} = \frac{1}{-2}.\frac{1}{1 - \frac{x}{2}}$

$= -\frac{1}{2} - \frac{x}{4} - \frac{x^2}{8} - \frac{x^3}{16} \cdots$; $-2 < x < 2$

6 **a** $\operatorname{cosec} x + \operatorname{cosec} x \cot x + \operatorname{cosec} x \cot^2 x + \operatorname{cosec} x \cot^3 x + \operatorname{cosec} x \cot^4 x \cdots$; $-1 < \cot x < 1$

b $\sec^2 x(1 + \tan^2 x + \tan^4 x + \tan^6 x + \tan^8 x + \tan^{10} x + \cdots)$ and since $\sec^2 x = 1 + \tan^2 x$, this expands as $(1 + \tan^2 x + \tan^4 x + \tan^6 x + \tan^8 x + \cdots) + (\tan^2 x + \tan^4 x + \tan^6 x + \tan^8 x + \cdots) = 1 + 2\tan^2 x + 2\tan^4 x + 2\tan^6 x + 2\tan^8 x + \cdots$

Answers

c similarly:
$(1 - \tan^2 x + \tan^4 x - \tan^6 x + \tan^8 x - \cdots)$
$+ (\tan^2 x - \tan^4 x + \tan^6 x - \tan^8 x + \cdots) = 1$

7 a $2, 2\frac{1}{4}, 2\frac{10}{27}, 2\frac{113}{256}$

b $1 + \frac{n}{n} + \frac{n(n-1)}{2n^2} + \frac{n(n-1)(n-2)}{3!n^3} + \frac{n(n-1)(n-2)(n-3)}{4!n^4} + \cdots$

c (i) 1 (ii) 1 (iii) 1 (iv) 1

d 1

e/f/g follow from above

h (i) $1 + x + \frac{x^2}{2} + \frac{x^3}{6} + \frac{x^4}{24} + \cdots$

(ii) relate to e^x

(iii) the derivative of the function is the same as the function

(iv) $= e^x$

Exercise 8 (page 137)

1 a $1 + 2 + 3 + 4 + 5$

b $1^2 + 2^2 + 3^2 + 4^2 = 1 + 4 + 9 + 16$

c $0^3 + 1^3 + 2^3 + 3^3 = 0 + 1 + 8 + 27$

d $0! + 1! + 2! + 3! + 4! = 1 + 1 + 2 + 6 + 24$

e $(-1)^1 + (-1)^2 + (-1)^3 + (-1)^4 + (-1)^5 + (-1)^6$
$= -1 + 1 - 1 + 1 - 1 + 1$

f $(-1)^1 1^2 + (-1)^2 2^2 + (-1)^3 3^2 + (-1)^4 4^2 + (-1)^5 5^2$
$= -1 + 4 - 9 + 16 - 25$ alternating signs (odd terms negative)

g $(-1)^2 + (-1)^3 + (-1)^4 + (-1)^5 + (-1)^6 + (-1)^7$
$= 1 - 1 + 1 - 1 + 1 - 1$ (odd terms positive)

h $(2 + 3) + (4 + 3) + (6 + 3) + (8 + 3) + (10 + 3)$

2 a 45 b 80 c 45 d 20

3 a $\sum_{r=0}^{5} x^r$ b $\sum_{r=0}^{5} (-1)^r x^r$

c $\sum_{r=0}^{4} (2x)^r$ d $\sum_{r=1}^{6} (-1)^{r+1} r$

4 a $3\sum_{r=1}^{n} r + 2\sum_{r=1}^{n} 1 = \frac{3n}{2}(n+1) + 2n$

b $4\sum_{r=1}^{n} r + \sum_{r=1}^{n} 1 = 2n(n+1) + n$

c $5\sum_{r=1}^{n} r - 3\sum_{r=1}^{n} 1 = \frac{5n}{2}(n+1) - 3n$

d $-6\sum_{r=1}^{n} r + 4\sum_{r=1}^{n} 1 = -3n(n+1) + 4n$

e $-2\sum_{r=1}^{n} r + 5\sum_{r=1}^{n} 1 = -n(n+1) + 5n$

5 a $\sum_{r=1}^{20} 4r + 8$ b $\sum_{r=1}^{10} 10r + 8$

c $\sum_{r=1}^{16} 30 - 5r$ d $\sum_{r=1}^{11} (22 - 3r)$

e $\sum_{r=1}^{9} 0.1r + 1.5$ f $\sum_{r=1}^{80} \frac{1}{4}r + 2$

6 b (i) $\sum_{r=1}^{11} (-1)^r (8r - 3)$ (ii) $\sum_{r=1}^{12} (-1)^r (9r - 2)$

(iii) $\sum_{r=1}^{10} (-1)^r (110 - 10r)$

(iv) $\sum_{r=1}^{19} (-1)^r (30 - 4r)$

c (i) $\sum_{r=1}^{16} (-1)^{r+1} (r + 3)$

(ii) $\sum_{r=1}^{12} (-1)^{r+1} (39r - 31)$

(iii) $\sum_{r=1}^{13} (-1)^{r+1} (302 - 17r)$

(iv) $\sum_{r=1}^{17} (-1)^{r+1} (63 - 2r)$

7 a $\sum_{r=1}^{8} 2 \times 3^{r-1}$ b $\sum_{r=1}^{11} 3072 \times \left(\frac{1}{2}\right)^{r-1}$

c $\sum_{r=1}^{7} 2000 \times \left(\frac{1}{10}\right)^{r-1}$ d $\sum_{r=1}^{7} \frac{9}{4} \times \left(\frac{2}{3}\right)^{r-1}$

Review (page 139)

1 a 11, 35, 107, 323

b $32k + p$, $32k^2 + kp + p$, $32k^3 + k^2p + kp + p$, $32k^4 + k^3p + k^2p + kp + p$

2 a $-\frac{1}{4}$, unstable b 12, stable

3 a $4n - 1$ b 23rd term

4 a $a = 3$, $d = -2$ b 20

5 132

6 277.5

7 $a = 2$, $d = 3$

8 17

9 a $u_n = 4 \times 3^{n-1}$ b 6th

10 3

11 390 625

12 a 292 968 b 533.3 (1 d.p.)

13 48.0 (1 d.p.)

14 a 9th b $|r| < 1$; 40

15 a $1 + x + x^2 + x^3 + x^4 + x^5 + \cdots$

b $\frac{1}{2} + \frac{3x}{4} + \frac{9x^2}{8} + \frac{27x^3}{16} + \cdots$

16 a e b e^3

17 $\frac{3n}{2}(n+1) + 2n$

Answers

Index